THE GREAT AUK

Some other books by Tim Birkhead

Promiscuity
The Red Canary
The Wisdom of Birds
Bird Sense
The Most Perfect Thing
The Wonderful Mr Willughby
What it's Like to be a Bird
Birds and Us

THE GREAT AUK

*Its Extraordinary Life, Hideous Death
and Mysterious Afterlife*

Tim Birkhead

BLOOMSBURY SIGMA
LONDON · OXFORD · NEW YORK · NEW DELHI · SYDNEY

BLOOMSBURY SIGMA
Bloomsbury Publishing Plc
50 Bedford Square, London, WC1B 3DP, UK
Bloomsbury Publishing Ireland Limited
29 Earlsfort Terrace, Dublin 2, D02 AY28, Ireland

BLOOMSBURY, BLOOMSBURY SIGMA and the Bloomsbury Sigma logo are
trademarks of Bloomsbury Publishing Plc

First published in the United Kingdom in 2025

A catalogue record for this book is available from the British Library

Library of Congress Cataloguing-in-Publication data has been applied for

ISBN: HB: 978-1-3994-1574-3; eBook: 978-1-3994-1573-6

2 4 6 8 10 9 7 5 3 1

Typeset by Deanta Global Publishing Services, Chennai, India
Printed and bound in Great Britain by CPI Group (UK) Ltd, Croydon CR0 4YY

Map of great auk range by Marc Dando

Bloomsbury Sigma, Book Eighty-Four

MIX
Paper | Supporting
responsible forestry
FSC® C171272
FSC
www.fsc.org

To find out more about our authors and books visit www.bloomsbury.com
and sign up for our newsletters

For product safety related questions contact productsafety@bloomsbury.com

CONTENTS

For Bob Montgomerie
Friend, collaborator, travelling companion and great auk enthusiast

Prologue

The great auk ruled the North Atlantic waves for millennia before being extirpated in 1844. Unlike many other human-exterminated birds, such as the dodo (1693), the passenger pigeon (1914) and the Hawaiian Kaua'i 'Ō'ō (around 1987), extinction has bestowed a compelling and enduring afterlife on the great auk.

The great auk's cachet reaches far back in time. Why else, 19,000 years ago, would people in southern Europe have adorned cave walls with life-size great auk images? Why else, 4,000 years ago, would the Maritime Archaic people of North America have interred their dead alongside great auk beaks, effigies and gizzard stones? Why else would someone in the Outer Hebrides around 400 BC have placed an entire head of a great auk within the walls of a house they were building? And why else would European seafarers encountering the great auk for the first time in the 1500s and 1600s have imagined it to be a witch?

For these past peoples the great auk was large, powerful and above all, magisterial in its human-like upright stance. For us today, the enduring aspect of the great auk is its flightless vulnerability and the quiet dignity with which it met its end, slaughtered by mindless, greedy men. The great auk embodies all those bird and other animal species that have been lost, or are vulnerable to being lost, at the hands of humans.

Even though the great auk itself no longer exists, its stuffed skins, empty eggshells and disarticulated bones continue to exert a powerful magic. Like the relics of saints, they are talismanic, objects of awe — pieces of the true auk — imbuing those that own them with an uncanny power. The great auk is a bird with a past, a present and, remarkably, a future. Why else would I be writing this book?

Just 75 eggs and 78 skins survive, and their rarity means that both museums and private collectors set great store by them. For almost two centuries, these fragments of a life once lived have been avidly sought and bought, often for exorbitant sums. The acquisition of great auk relics is, for some, an all-absorbing – and often self-destructive – obsession.

It is May 2017, and I am chatting on the phone to a retired miner and former egg-collector. In his strong South Yorkshire accent, he casually tells me that a mate of his is repairing a great auk egg. I'm stunned. With so few known eggs of this vanished, revered bird, it seemed extraordinary that not far away – and in an area of social deprivation made famous by the film *Kes* – such an egg exists. Amazed and intrigued, I ask who it is. My ex-miner tells me his mate's name, and agrees to introduce me. A few days later I go to his house to meet him. Once a miner too, and now a night watchman in the local steelworks, Graham Axon's hobby is creating replica eggs and repairing real ones.

I ask where the egg had come from, but Graham is evasive; 'I can't tell you,' he says. I am not surprised, for everything to do with great auk eggs is cloaked in mystery and intrigue, so I don't persist. When Graham says that the now completely restored egg has gone back to its owner my heart sinks. But he tells me that he has recorded his repair in photographs on his phone. We go into the shed in his garden, where Graham sits at his workbench and pulls out his mobile phone. In more detail than I had dared to hope, he reveals the entire process, from the arrival of the damaged egg, through its dismantling and the careful and seemingly miraculous reassembly of the 36 separate pieces. In a marvellous reversal of the fate of most eggs, Graham *was* able to put Humpty together again. The finished article is exquisite, and I feel sure that its owner, whoever it is, must be delighted.[1]

There was so much I wanted to know: who is the owner? How did they get the egg? How much had they paid to acquire it? How much had Graham been paid for his extraordinary repair work?

Among the paints, brushes and tools on Graham's workbench lay a curious, star-like cardboard structure that he picks up and hands to me. 'This was what I found inside the egg when I opened it,' he says. I am intrigued: the structure is an armature used to hold the fractured shell in place. Remarkably, there was handwriting on the cardboard, which might perhaps provide a clue to the egg's provenance. Graham told me that the owner wasn't interested in the armature and that I could probably have it. I was excited, since I knew that it would sit nicely in my university's Zoological Museum – a cabinet of curiosities of which I was honorary curator. All in all, it was a momentous day for me; a day that triggered a cascade of ideas and questions.

Throughout life, discovery has been my goal – finding something new, something previously unknown. The quest for new knowledge defines an academic career. As my academic colleague and friend Geoff Parker has so aptly said: 'All that a scientist should ever wish is to understand the natural world a little better.'[2]

A better understanding of the natural world is defined by discovery, and being first to discover something defines success in science. Yet almost by definition, discovery is elusive. Some discoveries are trivial, and many a scientist has had to accept – despite their best efforts – that making only minor discoveries has been their fate. The fortunate few strike gold and are lauded for it. Darwin's natural selection, Watson, Crick and Franklin's structure of DNA, Alec Jeffreys's DNA fingerprinting, Jennifer Doudna and Emmanuelle Charpentier's CRISPR, and so on. But discovery isn't simply made through luck or chance. As Louis Pasteur (of vaccination fame) said: 'In the fields of observation, chance favours only the prepared mind' – no less true today than it was when he said it in 1854.

The way that researchers make discoveries has intrigued scientists and philosophers for centuries. After returning from his round-the-world *Beagle* voyage with a wealth of personal experiences, Darwin used two strategies to fuel his creativity and clarify his thoughts. The first was to make a constant stream of observations coupled with ideas arising from a vast and varied panoply of correspondents, including other scientists, gardeners, bird breeders, pigeon fanciers and bee-keepers. The second was to take several circuits of his 'sandwalk' – a quarter-mile path on his property – each day, to provide himself with time to process all this incoming information. Thinking and walking. No one pushed the envelope like Darwin. More than anyone else, Darwin's 'pushes' punctured the all-enveloping dogmas of the Christian church to redefine how we understand the natural world.

Crossing boundaries and trying to think outside the box is what I have done as part of my own scientific endeavours. An interest in the history of ornithology, and how we know what we know about birds, inspired me to broaden my horizons by speaking to people outside science and outside mainstream ornithology – like Graham.

And here I was, in his shed, standing on another bridge into the unknown. Thirty years previously I had written about the great auk in a book I called *Great Auk Islands*. The title is a double entendre, recounting my discoveries and adventures studying auks, mainly common guillemots – a second cousin to the great auk – on some great North Atlantic islands. Like almost everyone who studies auks, I was irrevocably drawn to the ultimate auk: the aptly named great auk, which sadly is no more.

While writing *Great Auk Islands*, I was surprised to learn that several of the known great auk eggs were missing, their whereabouts a mystery. These were eggs that had once been part of several private collections, and each was worth tens or hundreds of thousands of pounds. They were the zoological equivalent of the lost paintings of Dutch artist Johannes Vermeer. Intrigued, I began to ask about those missing eggs, but no sooner had I started than I came up against Bill Bourne, the legendary and irascible old man of seabird biology, who told me to mind my own business. Reluctantly, I did.[3]

Nonetheless, my obsession with the great auk remained undiminished and, in the years that followed, I eagerly read each new account of this extraordinary bird as it appeared. The most comprehensive of these was a beautifully produced monograph by Errol Fuller. Privately published in 1999, it lists all known specimens of mounted skins, skeletons and eggs, and was to become my great auk handbook. Initially, though, I was disappointed, since as a biologist, my interests were focused more on what the bird had been like while alive rather than on its relics. I realised later that Fuller's book was essentially a collection, in this case of information. His aim was to provide the most comprehensive version of an exercise that started soon after the bird became extinct, cataloguing the history and whereabouts of every great auk skin, skeleton and egg. The need to continually update these catalogues reflects the fact that great auk relics are moving targets. They are part of the constantly changing cycle of acquisition and disposal as owners die, and relatives discard or sell their belongings. Occasionally, as Fuller describes, previously unknown great auk relics appear on the market, although with time such discoveries grow less and less common.

When I got home, I went straight to Fuller's book in the hope of identifying the egg Graham had so skilfully repaired. And there it was! Like all of the 75 known great auk eggs,

Fuller had given this one a name: Bourman Labrey's Egg. There were no details of who Bourman Labrey was, other than the fact that he once owned the egg, and lived in Manchester during the mid-1800s. As I later discovered, that name was the result of a misreading. His name was Beebee *Bowman* Labrey, not Bourman, so this is now the Bowman Labrey Egg.

What made me so certain that this was the egg that Graham had repaired was the fact that Fuller states that the egg had been broken a long time ago and in the 1840s had been 'crudely repaired' by one of Britain's leading ornithologists, William Yarrell. And indeed, when I compared the handwriting on the egg armature I had seen on Graham's workbench with Yarrell's handwriting, they matched.[4]

The damaged Bowman Labrey great auk egg, repaired by William Yarrell and photographed in the 1890s by Edward Bidwell.[5] The egg was in an even worse state when Graham Axon received it for repair in 2017. (From Tomkinson & Tomkinson, 1966)

There was more. To my amazement, this was one of the missing great auk eggs. It was one of 13 once owned by an eccentric millionaire – Captain Vivian Hewitt – that had disappeared after his death in 1965. Perhaps now, 30 years after I first started to think about them, here was an opportunity for me to discover the fate of those missing great auk eggs.

The great auk's extinction in 1844 coincided with a booming interest in natural history. Acquiring specimens of mollusc shells, ferns, birds' eggs and stuffed skins, often under the shady umbrella of science, provided opportunities for men (and it was mainly men) to compete. It allowed them to show off and sometimes to contribute to the growing body of natural history knowledge. Collecting became an obsession. Personal collections were a human manifestation of Charles Darwin's idea of sexual selection. Collections were like the candelabra racks of antlers and the gaudy plumes that had evolved in male deer and pheasants, for example, to compete for females, or to make the owners more attractive to them. To render his idea of sexual selection accessible to a general readership, Darwin used the current and popular practice of artificial selection of domestic animals as an analogy. Like its sexual counterpart, artificial selection created animals of extreme appearance, including huge-rumped cattle, ultra-sleek running dogs and fan-tailed or feather-footed pigeons. Darwin could equally well have used the human habit of extreme collecting as a metaphor for sexual selection, for this too was familiar territory for everyone at that time.

Victorians grew up amid these various competitive activities. For animal breeders, country shows provided ritualised opportunities for displaying the fruits of their labours. There were cat clubs and dog clubs, as well as poultry, pigeon and

canary clubs. In contrast, collectors of natural history
specimens had fewer public opportunities to show off their
treasures. Most of their 'displays' were private affairs conducted
in the seclusion of their studies and salons. Indeed, for those
who collected birds' eggs, keeping them in the dark in specially
constructed cabinets was essential if the eggs were to retain
their colours and vitality. Collectors formed cliques in which
participants knew, or knew of, each other, with those that
collected birds' eggs and stuffed birds being the largest of such
communities. Birds were abundant in the 1800s and rather
little skill was needed, at least for common species, to find,
clean out and display their eggs. For over a century, country
boys collected eggs. Most grew out of it, but for the few that
didn't, their harmless hobby became a life-long obsession.
Adult egg-collectors called themselves 'oologists' and their
hobby 'oology', justifying it as science. Stuffed birds required
rather more skill to prepare, but there was no shortage of
taxidermists capable and willing to bring shot specimens back
to a semblance of life under a glass dome.

Although illegal today, the hobby of collecting birds' eggs
or skins in the 1800s and first half of the 1900s was completely
normal. Indeed, collectors comprised much of the community
of natural history enthusiasts. As with all social groupings,
there were hierarchies. The lowest in the oological pecking
order were the boys who collected eggs for fun. Next were
those who systematically collected just one egg from the
nests of common bird species. Then, there were those who
coveted entire clutches of eggs, and especially those of rare
species. Numbers mattered; larger collections meant more
kudos, as did the eggs of more exotic species. Many collectors
specialised in the eggs of rare birds such as eagles and falcons,
or in those that laid particularly beautiful eggs, like the
common guillemot. At the very top of the hierarchy were
those who acquired one or more eggs or mounted skins of

birds that were no longer with us, like Madagascar's elephant bird, New Zealand's huia, the North American passenger pigeon and, of course, the great auk.

The great auk's egg is large, beautifully shaped and exquisitely marked. It fits perfectly in one's hand. They are extremely rare. Acquiring one required exceptional wealth or extraordinary good luck. Empty eggshells, and the birds' emptied and stuffed skins, were often bought and sold for more than their weight in gold.

Those writing about birds in the early 1800s — and there were many — felt obliged to include an image, a description and some comments on every species. The focus then was very much on what the bird and its eggs looked like. This was the era of 'descriptive ornithology' — a vast cataloguing exercise of extant and extinct birds. The aim was to position all bird species in their appropriate position on the tree of life. It was to be more than a century before that exercise neared completion and the tree started to bear fruit. Creating this great ornithological catalogue sounds innocent enough, but it meant killing birds, and in large numbers, for this was an exercise that could be undertaken only through the acquisition and examination of specimens — the feathered skins of dead birds and their empty eggshells. It was an era in ornithology when birds were dying for knowledge. Initially amassed by private individuals, many of these specimens eventually ended up in national museums, where they can be found today.

The great auk — and specifically, its mortal remains — emerged phoenix-like, at the centre of a social network that connected many of the nineteenth and early twentieth century's great ornithologists. This was a network that comprised individuals whose interests often lay more with the trappings of wealth than with ornithology. It also included those who would go to almost any lengths to wrap their fingers around a great auk egg or skin.

Vivian Hewitt on Puffin Island near Anglesey in the 1920s. (Courtesy the Hewitt Papers)

The most extraordinary of these obsessives was Vivian Hewitt, whose vast wealth allowed him to accumulate more great auk remains than anyone else: 13 eggs and 4 mounted skins.[7] Those eggs, like all the others that still exist, are part of a biological bloodline. They are also fragments of another dynasty, an empire of ownership that stretches back to a moment in the late 1700s or early 1800s when a seafarer on a remote, wave-washed Atlantic island reached down to pick a great auk egg up from the bare rock surface on which it had been laid. It sounds pretty harmless, but great auk eggs and skins were trophies acquired only in the aftermath of a terrible slaughter, in which parent birds were killed – for food, for feathers or, towards the end, to be turned into scientific specimens until they were no more. Thumbing through an art book as a small child, I came across Pieter Bruegel's painting *Massacre of the Innocents.* Its depiction of human atrocities is an image that has haunted me ever since, even though much of the original horror had long been painted out, just as it has been with the slaughter of great auks.[8]

A few tens of great auk eggs and mounted skins, together with the written records of those involved in its extirpation, are all that now exists to allow us to remember this magnificent and unusual flightless bird. My aim has been to use these relics, together with my knowledge of the lives of extant auks, to better unravel some of the great mysteries surrounding the great auk – not only how it lived, but how it died. What was so special about this bird? What caused it to go extinct before any ornithologist ever saw one alive? What lessons can we learn about ourselves as the collectors and later as the protectors of great auk specimens, and as the perpetrators of the current precipitous global decline of bird populations?

Having discovered the provenance of the specimen that Graham repaired in 2017, I thought I had identified a crack in the oological underworld that would allow me to discover what happened to the other missing great auk eggs. But it turned out there was more secrecy to penetrate.

There things stood until 2021 when, late one evening, my phone rang and someone I had a vague recollection of meeting long ago introduced himself. He told me he was clearing the house of a recently deceased friend, had found some guillemot eggs, and wondered if I'd like them. I was thrilled. I had been doing research on guillemot eggs for the previous decade, trying to understand why they are such an extraordinary shape. Now I was being offered a collection of eggs that would make an interesting addition to our university museum.

I accepted, of course, and a couple of weeks later drove the hundred or so miles to pick up the eggs. They were among the most exquisite I had ever seen. Over coffee, surrounded by vast numbers of books and bird specimens in the house he was clearing, I mentioned to my host the similarity between guillemot eggs and great auk eggs. I also told him that for 30 years I had been intrigued by Vivian Hewitt's missing great auk eggs.

'I think I might be able to help you there,' he said. The hairs on the back of my neck bristled with excitement. 'Yes,' he said, 'I can tell you the story.' And in due course, he did.

This book grew from that story. It is in two parts. The first is about the living bird and what made it so special. The second is its afterlife following its extinction, a tale of human folly and our disregard for the natural world.

PART 1
LIFE

Chapter 1

Funk Heaven

Waves of contractions sweep along the great auk's uterus. She is standing up on her webbed feet in an unusual posture. Her head has rolled back and sunk into her neck, her eyes are closed and her tiny wings droop by her side. She shudders with each contraction. The conspicuous bulge on her belly and just above her tail signals the imminent arrival of a huge egg. Standing close by, the female's partner seems alternately indifferent and inappropriately solicitous. His untimely attempts to preen her head are rejected with a twitch of irritation. After 10 minutes, she is still standing upright and beneath her tail the pointed end of the egg is just visible. More uncomfortable minutes pass and then, with one huge heave, the rest of the egg emerges in a bulbous glistening rush. Both partners stare down at it, taking in its distinctive colours and markings. Almost immediately the female leans forward and, guiding the egg with her beak, pushes it under one wing, and then falls gently forward onto her breast and begins to incubate.

At almost one metre (3ft) tall, sporting a pair of conspicuous spectacles and the avian equivalent of a dinner jacket over a white shirt, the great auk – both male and female – in its breeding attire is a stately and majestic bird. It is flightless, but underwater its tiny wings become flippers endowing the bird with the agility of a fish – actually, greater agility than a fish, since it can out-swim most of the smaller ones. This is a bird that through its large body size and flightless wings is supremely adapted to exploit the ocean's rich inshore waters.

Our great auk couple is not alone. All around there are thousands, or perhaps hundreds of thousands, of others. They are packed together on a tiny, low-lying slab of granite, 65 kilometres (45 miles) off Newfoundland's north-east corner, just 25 hectares in extent. This is Funk Island, and the year is 1400. The year is important, for apart from the very occasional visit by indigenous peoples and the odd polar bear, the birds breeding here live in a world that is both peaceful and almost entirely their own. Actually, calling it peaceful is both right and wrong. Compared with what later generations of great auks endured, life on the Funks (as the island and two nearby rock shoals are known) at this time is serene: an auk heaven on earth. That said, this vast city of auks is actually anything *but* peaceful. It is a raucous hubbub of activity, busy and bursting with life during the warm but brief summer months. Only at night, under the ethereal light of the Milky Way and the occasional aurora borealis, is the colony quiet.

Three weeks before our female produced her egg, Funk Island was still ice-bound. Vast numbers of great auks had swum north from their milder, ice-free wintering grounds, along the eastern seaboard of present-day America. The sea-ice barrier brought the birds to an abrupt halt as they approached Funk Island. The great auk's cousins, razorbills and guillemots, simply took wing and flew over the ice to the island.

Milling about the ice-edge, the great auks swim to and fro, waiting. As the air warms over the next few days, the ice begins to fracture, and jagged black leads like cracks in an eggshell begin to appear. The auks surge forward following the break-up of the ice until finally – after nine months at sea – they touch land again. Falling forward through the swell that washes up the western end of the island, they use their rough-soled feet and sharp claws to grip the granite and emerge upright from the waves. Waddling like a stream of tight-rope walkers, they head towards the centre of the island, their diminutive wings outstretched to help them balance. With an uncanny spatial

awareness, each bird walks back to the exact spot on the bare
rock, just 30 by 30 centimetres (12 inches) square, where they
bred last year. It is probably where some females have laid their
egg for the past 30 years. Simply being there announces
ownership of this tiny, otherwise featureless spot. Others that
come too close are repulsed by an outstretched neck, an open
beak and a growl. Within days most of the birds have returned;
their numbers are vast, the noise relentless. Were any human to
be present, the stench of the birds' droppings would sear their
senses. The rock that was once pinkish granite is now a Jackson
Pollock splattering of black and white, interspersed with the
brilliant yellow gapes of displaying auks.

On their return to the island, each bird has just one thing
on its mind. For the male – predictably – it is sex. For the
female it is her egg – making it, laying it and protecting it.
And, of course, both partners are preoccupied with one
another. They have bred together for years, and after a winter
apart are excited to be together again. Like most other
members of the auk family, great auks are socially monogamous.
Partners work together to incubate their single egg and, all
being well, rear their offspring to independence.

Stepping out of the waves, our male ploughs through the
antsy crowd of conspecifics towards his breeding site. His
partner has seen and recognised him from some distance, and
stands up ready to greet him. Her head feathers are raised in
excitement, as are his, and holding her head downwards she
opens her beak and releases a deep guttural growl. Stepping
forward he wraps his head and neck around hers and together
they rumble and purr in an ecstatic tactile embrace. They
inhale each other's deep, musky odour, much as humans do.
She drops suddenly forward into a prone position, and uttering
a mechanical grinding call, throws her head back and raises
her tail. He seems to be taken by surprise, but quickly recovers
and steps smartly onto her back with his little wings flapping.

He lowers his tail to perform a protracted cloacal kiss, transferring some of the sperm that might in due course fertilise her single ovum. But not yet. For now, those precious male sperm cells worm their way into special sperm storage tubules within her oviduct. They will remain there, safe and viable, for several weeks, before – when the time comes – swimming to the waiting ovum higher in the female's reproductive tract.

All around, other recently reunited pairs are preening each other, displaying and copulating. This is the noisiest and liveliest part of breeding season. The startling black-and-white plumage of this great mass of birds shimmers against the azure sea beyond.

The egg is large, almost 14 centimetres (5½ inches) long and 9 centimetres (3½ inches) wide, and extremely pointed at one end. Our female's egg is cream-coloured and decorated with an intricate mesh of random squiggles, as though drawn on with a black felt-tip pen. The thick eggshell is rough to the touch, providing vital tactile feedback to the parent's brood patch as it is incubated on the bare rock. Like their guillemot cousins, great auks make no nest.

Both partners take turns to incubate, with the off-duty bird usually out at sea feeding, or, if not, standing nearby and gently preening its partner – an act known as allopreening. But there's something else going on. A few metres away from the breeding area on elevated, uneven terrain unsuitable for egg-laying, off-duty birds assemble in what seabird researchers call a 'club'. Away from their parental duties, birds spend time here as though simply watching the world go by. With her partner now taking his turn on the egg, our female auk scrambles up to join the club. Suddenly, she is confronted by a challenging male. He's more ardent than aggressive, but he's trying it on, offering to allopreen her. She rejects his advances – she's already laid her egg, after all – but other females don't or cannot resist. This is a club where extra-marital relations are formed, and illicit sex occurs. And, most remarkably, some of these females – those yet to lay an egg – seem to be especially inclined towards one particular male. This unexpected arrangement – one that also occurs in their close cousin, the razorbill – is an opportunity where, out of sight of their partners, females can seek better genes for their offspring. Such illicit liaisons may also harbour longer-term ambitions. They may be like an insurance policy in the event of their partner failing to return one spring – a relationship half-formed but ready to go.[1]

Incapable of flight, the great auk was much larger than the other auks. At 3.6kg (7.8lb), it could dive deeper and remain submerged for longer than guillemots (1kg, 2.2lb) or razorbills (0.6kg, 1.3lb), its smaller cousins. Its size and huge bill allowed it to capture fish such as full-grown herring, which were much too large for the other auks that might otherwise have been competitors. Its long, laterally compressed, blade–like beak is one of the great auk's most distinctive features. Both mandibles bear characteristic corrugations. On the upper bill the four or five furrows are long and curved; on the lower, the six to ten grooves are short and straight, like knife cuts. As in the puffin and razorbill, young birds' bills bear fewer grooves and it is likely that they signalled a bird's age and were valuable in the business of selecting a partner. Some people have wondered whether one or more of the great auk's grooves were once white, as they are in the razorbill, suggesting that perhaps the white pigment has been lost over time. I doubt it, for the razorbill's white furrows are extremely robust and an integral part of the beak covering.[2]

There were an awful lot of great auks on Funk Island in 1400, perhaps as many as a quarter of a million pairs. How and where, one wonders, did they all find enough fish to sustain themselves and their growing chicks? The other auks – razorbills, puffins and guillemots – fly up to 60 kilometres (around 40 miles) from the colony, and almost certainly find areas of good feeding by retracing the flightlines of airborne commuters returning to the colony. Great auks had no such obvious or convenient source of information. But maybe they didn't need it. Everything we know about Newfoundland's seas before the arrival of Europeans tells us that fish were superabundant – more abundant than we can imagine today. In 1400, the marine environment was pristine. Nutrients welled up from the deep to fuel the phytoplankton that in turn fed the zooplankton that provided the sustenance for the capelin, herring and cod that filled the seas.

The way in which the flightless great auk found its food beneath the sea's surface remains a mystery. Even for the well-studied southern hemisphere penguins, we have little idea how a bird that forages in the ocean's dark depths finds its prey. It can be hard to imagine what it is like to be a bird. It once seemed impossible that certain seabirds might use their sense of smell to find feeding areas, but we now know that this is exactly what albatrosses and petrels do. Might great auks have done the same? Perhaps they possessed senses we currently cannot imagine? Until a few years ago, who would have thought that birds can see colours that are invisible to us; that white storks could detect the distant aroma of new-mown hay; that honeyguides use the scent of wax to locate bees' nests; or that birds could sense the earth's magnetic field to find their way? Great auks may have used their hearing to home in on the sounds we now know that some fish make.[3]

The inability to fly is the great auk's most distinguishing feature, and the one that would be its undoing. Why did it abandon flight? The short answer is energetics. The largest flying auks today are the common guillemot and Brünnich's guillemot, both of which weigh around 1kg (2.2lb), and like all other auks they use their wings when diving to propel themselves underwater. Wings that are large enough to keep a guillemot airborne are too big for submarine flight in the denser medium of water, unless they are folded to make them smaller. This is what guillemots and other auks do. The size of the wings of flying auks is therefore a compromise between aerial and submarine flight. If the guillemot or any other auk was heavier it would need disproportionately large wings to be able to fly, incurring unsupportable energetic costs, and the wings would also be too big to reduce by folding when underwater.

In terms of diving, a larger and heavier body is much more efficient, increasing both the depth to which a bird can dive and its overall energetic efficiency. There is compelling evidence for this energetic scenario, since among both extant and fossil wing-propelled pursuit-diving birds, all of those weighing more than 1kg are flightless, exemplified by the fact that both the smallest penguin species (the little blue) and the largest flying auks (the guillemots) both weigh about 1kg.

Sacrificing the ability to fly allowed the great auk to evolve a larger, heavier body and become a super-efficient underwater predator.[4]

I have spent most of my career studying guillemots, mainly on Skomer Island off the coast of Wales, but also in Newfoundland and Labrador. In truth, most of my time was spent teaching undergraduates (which I thoroughly enjoyed), applying for research grants and wrestling with an increasingly burdensome university bureaucracy. Guillemots on Skomer were the subject of my D.Phil. from 1972 to 1976 and I have kept the study going ever since. Here I am now, 50 years on and still intrigued by auks. Over my career, I have investigated the biology and behaviour of a wide range of birds in different parts of the world, often with research students, but I have probably spent more time watching guillemots than any other species. I have also studied razorbills and Atlantic puffins, and, for several years, the common guillemot's northern counterpart, the Brünnich's guillemot in the Canadian High Arctic. Half a century gives one a sense of knowing these Atlantic auks quite well, but I take nothing for granted and they continue to surprise me. With the recent impact of climate change and avian flu, nothing remains the same for long.

My D.Phil. was essentially 'ecological', designed to investigate the population biology of guillemots. Along with puffins and razorbills, guillemot numbers in southern Britain were in steep decline in the 1970s, and my task was to find the cause. However, my main motivation for studying birds has always been to learn more about their behaviour. In the early 1970s, ecology and behaviour were distinct disciplines and within the Zoology Department in Oxford, where I conducted my D.Phil., these subjects were the domain of different research groups. The Animal Behaviour Research Group was headed by the world-famous Niko Tinbergen, who won the Nobel Prize in Physiology and Medicine in 1973 for his pioneering research. The research group I was part of was the Edward Grey Institute of Field Ornithology, whose equally eminent director, David Lack, was an expert in the ecology of birds, focusing on understanding how their populations functioned.

Luckily for me I was free to study both the ecology and behaviour of guillemots, and, even more fortuitously, my time as a research student coincided with the emergence of a new discipline, 'behavioural ecology'. This was a merging of two subject areas – a new way of looking at how animals, including birds, lived. Behavioural ecology was the study of how behaviour played out on an ecological stage, observed through the lens of natural selection. The supreme novelty of this approach was that this was natural selection operating explicitly at the level of individuals. Behavioural ecology launched a new and massively productive era of ornithological discovery. It may be no coincidence that David Lack, probably the twentieth century's most famous ornithologist, had long been a firm advocate of individual selection and it was this that enabled him to understand how populations worked. Indeed, when Richard Dawkins published *The Selfish Gene* in 1976 – a popular account of behavioural ecology – he singled out David Lack as one of the pioneers whose thinking helped to shape this new field. Sadly, Lack did not live to see these particular fruits of his labours, for he died in 1973 at the very start of my studies.[5]

Studying the behaviour and ecology of guillemots and razorbills has allowed me to speculate here about how great auks would have lived. Almost all seabirds share a set of ecological traits, and many breed in vast aggregations – like the great auks at Funk Island – mainly on inaccessible offshore islands or on sheer cliffs where they are safe from terrestrial predators. Seabirds are long-lived and they are slow reproducers: most do not begin breeding until they are several years old. The majority of species lay and incubate only a single egg, because their food supply, being far out at sea, prevents them bringing sufficient food back to the colony to rear more than one chick at a time. In terms of behaviour, seabirds tend to be socially monogamous, forming a long-term relationship with the same breeding partner to breed year after year. The social glue that keeps partners together includes elaborate greeting

ceremonies, endless preening of each other's head-feathers
and regular copulation. Promiscuous mating and divorce
occur occasionally, and, if one partner dies, the other typically
finds a replacement rather quickly. Not that different from
humans, really.

After weeks of incubation, our great auk's egg hatches. The
newly emerged chick is brooded continuously, kept warm and
safe from the deadly beaks of predatory gulls. The parents take
turns to bring fish like capelin and herring, which the chick
swallows whole. The fish are small at first, but increasingly
larger as the chick grows. In what seems no time at all, it is
time for the chick to take to the sea, where it is cared for by its
father for the next two months. Amid thousands of others,
father and chick abandon their tiny breeding site at dusk.
Following traditional walkways across the island, they wend
their way towards the sea. There is a moment or two of
hesitation before father and offspring plunge into the waves
and swim out into the night. Paddling south-westwards they
pass Nova Scotia, surging on and on towards the Carolinas.
Others swim north, and some end up off the west coast of
Greenland. Most of the time they are out of sight of land. Fish
are abundant and the adult auk dives for food, returning to the
surface to stuff its offspring's squeaking mouth. A week or so
after leaving the colony, the male starts to shed what are
inappropriately referred to as his 'flight feathers', his primaries,
the longest wing feathers. As in most birds, the great auks lose
these feathers one by one over a few weeks.[6] This is different
from the razorbill and guillemots, which drop their flight
feathers simultaneously, and as a result are rendered flightless.
Since their recently fledged chicks are unable to fly, and
because adult razorbill or guillemot wings without primaries

are still perfectly good as underwater paddles, this is an excellent arrangement. For the great auk, whose flightless wings have evolved to be just the right size as underwater paddles, any further reduction in their size by the simultaneous loss of their primaries would be a huge disadvantage. By shedding its wing feathers one by one, the great auk has evolved a pattern of moult that breaks the auk mould, and is a clear adaptation to being flightless.

As the primary feathers are lost and new ones begin to grow, the rest of the great auk's body feathers are gradually lost and replaced too. The result is a transformation from a black-faced bird with white spectacles to a white-throated bird with plumage similar in appearance to that of a young great auk. We know this from the tiny handful of winter adult specimens that still exist.

In late August 1821, the cleric and naturalist John Fleming, travelling around the coast of Scotland, wrote:

> … *we got on board a live example of a Great Auk* (Alca impennis), *which Mr Maclellan … had captured sometime before, off St Kilda. It was emaciated, and had the appearance of being sickly; but, in the course of a few days, it became sprightly, having been plentifully supplied with fresh fish, and permitted occasionally to sport in the water, with a cord fastened to one of its legs, to prevent escape … A few white feathers were at this time making their appearance on the sides of its neck and throat, which increased considerably during the following week, and left no room for doubt, that, like its congeners* [other auk species], *the blackness of the throat feathers of summer is exchanged for white, during the winter season. I may add, that the black colour of the throat of the Razor-bill* (Alca torda), *was at this time undergoing a similar change.*[7]

Why the adult auks adopt a distinct and white winter plumage remains a mystery. One possibility is that the dark summer head-plumage of great auks, guillemots and razorbills signals something about their health and condition to existing or potential partners. I think this is unlikely, since, over the years, I have seen many breeding guillemots (that is, those with a mate) with white-flecked head and neck plumage as though they failed to complete their moult. Another idea is that the white winter plumage somehow makes the capture of fish and other prey easier. Certainly, the dorsal black and ventral white plumage pattern so typical of many seabirds is thought to be countershading that makes the birds less conspicuous to predators or to their own fishy prey. But, as with so much, we really do not know the answer to these questions.

By late August all the great auks and other seabirds have left Funk Island to spend the winter at sea. High waves break over parts of the island through much of the winter, washing some of the season's guano back into the surrounding seas. Dissolved, the seabirds' nitrate-rich droppings provide the nutrients on which phytoplankton depend, fuelling the super-rich marine ecosystem of which the great auk is monarch.

Within a couple of months in late winter or spring, the birds' winter attire is replaced by smart summer finery as they start to head back to the Funks once more. The cycle of nutrients, plankton, fish and birds has been running pretty much unchanged for millennia.

Along with other places where great auks once congregated to breed, Funk Island in the 1400s was an alcid idyll. It was an idyll that was soon to be shattered.[8]

Chapter 2

Funk Hell

The peace that had pervaded Funk Island for millennia was destroyed for ever in 1520 by the arrival of João Álvares Fagundes and his entourage.

Five hundred years earlier, Norsemen had traversed the North Atlantic and established settlements on both Greenland's south-west coast and the northern tip of Newfoundland. They undoubtedly enjoyed and exploited the region's abundant natural resources, including cod and seabirds, but they appear not to have found Funk Island.

Christopher Columbus's rediscovery of North America in 1492 launched an inrush of European explorers seeking riches and new opportunities. And riches there were, for as Basque fishermen had already discovered, this was a new world of unparalleled plenitude. Hearing of Newfoundland's treasures – the New Land of the Cod Fish – Fagundes set off from Portugal in 1520, collecting *en route* in the Azores a bunch of optimists hoping to settle on Cape Breton in Nova Scotia.

Some 40 miles from the coast of Newfoundland, Fagundes found further evidence of God's willingness to provide: an almost unimaginable abundance of great auks on Funk Island. After seeing, and most probably eating, some of the birds, Fagundes dubbed the island 'Isla de Pitiguoem' – island of penguins.[1] News of Funk as a fast-food stopover for transatlantic explorers spread rapidly. By 1527 European mariners, impatient for fresh meat after their long Atlantic crossing, were known to be feasting regularly on the island's great auks.[2]

In the spring of 1534, the French explorer Jacques Cartier approached Funk Island knowing full well what to expect: an island full of birds. 'In less than half an hour,' he wrote, 'we filled two boats of them as if they had been stones: so that besides them that we did eat fresh, every ship did powder and salt four or five barrels of them.' His less well-informed men, however, on seeing the unfamiliar, large, upright birds for the first time, thought they were witches, and were wary. Indifferent to his crew's concerns, Cartier organised a killing spree, his account of which was guaranteed to inspire and encourage others:

> *The Apponat* [the great auk] *whose numbers are so great as to be incredible, unless one has seen them; for although the island is about a league in circumference, it is so exceedingly full of birds that one would think they had been stowed there ... Some of these birds are as large as geese, being black and white with a beak like a crow's. They are always in the water, not being able to fly in the air, inasmuch as they have only small wings about the size of half one's hand, with which, however, they move as quickly along the water as the other birds fly through the air. And these birds are so fat that it is marvellous.*[3]

On a second trip, a year later, Cartier wrote:

> *The island* [Funk] *is so exceedingly full of birds that all the ships of France might load a cargo of them without perceiving that any of them had been removed.*[4]

So well-known did the island become that in 1536 an English part-time geographer, Richard Hore, made a point of stopping at Funk with 30 gentlemen tourists, to enjoy a great auk feast.[5]

Pretty helpless on land, beneath the waves the great auk was essentially a penguin, and long before true penguins had been discovered that was what the great auk was called. As Fagundes's account indicates, the origin of the name is the Portuguese term *pingüe*, meaning plump. *Pitiguoen* is what Fagundes called them, and they were indeed 'marvellously fat'. Sailors from what would soon become Spain called it *el Pájaro Bobo*, the silly bird, just as other easily killed birds were called 'boobies', 'loons' or 'dodos'. Jacques Cartier (and later, Captain Cook) called the great auk 'apponath' or 'arponaz', a name he and his men appropriated from Newfoundland's indigenous people, the Beothuk. When writing about his journey to St Kilda in 1698, Martin Martin referred to the great auk as the 'gairfowl' or 'gare-fowl', a name derived from the Icelandic 'geirfugl', in which 'geir' – 'spear' – makes reference to the bird's spear-like beak. It was the naturalist Thomas Pennant who, in 1776, introduced the name 'great auk', to distinguish it from two other, smaller auks, the razorbill (sometimes then called simply 'the auk') and the little auk.[6]

In their 1775 survey of the coasts of Newfoundland and Labrador, Captain James Cook and Michael Lane renamed Fagunde's 'Isla de Pitiguoem' Funk Island – a name that captured the foul odour of the birds' ammonia-rich droppings.

Captain Cook and Michael Lane's 1775 map of Newfoundland. I have added an arrow to show the location of Funk.[7] (Courtesy K. J. Korneski)

After his first voyage to the coast of North America in 1497, John Cabot wrote to his father, telling him that the seas 'yeeldeth plenty of fish, and those very great, as seales, and those which we commonly call salmons: there are soles also above a yard in length: but especially there is a great abundance of that kinde of fish which the Saluages ["savages" – indigenous people] call Baccalaos [cod].'[8] Once this abundance became known, a gold rush was inevitable. At the time of Fagundes's arrival, the seas around Newfoundland still abounded in cod, seals, whales and seabirds; by the 1580s there were 50 English, 50 Portuguese, 100 Spanish and 150 French vessels hunting cod off Newfoundland, many of them making use of Funk's meat store as they passed, marking the beginning of the end for the 'witches' that bred there.

The earliest known great auk fossils date back half a million years, and the first modern humans arrived in Europe around 40,000 years ago, so great auks had many millennia of relative peace and quiet.[9] They were not, however, without predators. Foxes and bears forced great auks and other seabird species to breed either on offshore islands or, for those that could fly, on inaccessible mainland cliffs. Biologists tend to think of Funk Island, with its gently sloping shores, as typifying the great auk's breeding habitat. But just look (on television or online) at the kinds of places where the equally flightless southern hemisphere penguins can hop, jump and scramble to breed. Indeed, the meticulous anatomical observations that the ornithologist Johann Naumann made in the 1840s of the 'breadth and roughness' of the soles of the great auk's feet enabled him to say that the bird climbs sloping surfaces more easily and dexterously than guillemots and razorbills. I am convinced, therefore, that before modern humans arrived on the scene, great auks occupied a wider range of breeding habitats and locations around the North Atlantic's shores. Evidence bursts from the numerous archaeological finds – not all, of course, from breeding colonies – but still, strongly suggestive of an extensive geographic range.[10]

As soon as humans and great auks collided, the birds and their eggs became extremely vulnerable. But human populations in the past were small, and great auks would have flourished even at some sites known to man. They were safe, even from polar bears, with whose geographic range the great auk's range barely overlapped. There are exceptions, of course. When Jacques Cartier landed on Funk Island in 1534, he was shocked to encounter a polar bear, left behind after the winter sea-ice had retreated.[11] That bear must have caused havoc among the great auks and other seabirds, not only through its predation of eggs and adult birds, but by the sheer pandemonium it created as the birds fled in panic and their eggs rolled away or were trampled. The bear got its just deserts – it was killed and eaten by Cartier and his crew.

As long as they were only occasional, encounters with predators like polar bears would have had a negligible effect on great auk numbers. The great auk's long, slow lifestyle meant that the occasional breeding failure had little impact on overall numbers, as it still does for guillemots and other seabirds.[12]

Over time, human populations began to increase, and as they did, they started to encounter great auks more often. The birds began to choose breeding locations that were inaccessible or unknown to humans as a result. In a way, it was a kind of arms race: human technology such as canoes or kayaks providing better access to breeding sites, and bows and barbed arrowheads increasing people's killing power, versus the ability of the birds to relocate to safer, more out-of-reach sites. In truth, it was more a case of advance and retreat: human advance, great auk withdrawal. As the ornithologist Henry Seebohm wrote in 1885 with inept anthropomorphism: 'It never appears to have entered into the calculations of the earlier generations of great auks that sooner or later evolution would produce a race of sailors to whom no flat coasts would be impregnable.'[13]

The human advance was slow but sure, and the relationship with great auks – despite being weighted in humanity's favour – endured for millennia. It is a story that can be pieced together from fragmentary, but often spectacular, archaeological evidence.

One of the earliest indications of our interactions with great auks is also among the most impressive. It comes from the 19,000-year-old images in the Cosquer Cave, near Marseilles on France's Mediterranean coast. The submarine entrance to this cave was discovered by the diver Henri Cosquer in 1985. One can only imagine his feelings as he surfaced inside the cave to see with his flashlight a vast gallery of human handprints, horses, ibex, seals and three great auks. I find just being in a cave with prehistoric art emotionally exciting, but for Cosquer the realisation that he had discovered something momentous and was the first modern person to see those images must have been truly extraordinary.[14]

Now rated in the same league as Lascaux, Altamira and Chauvet, Cosquer Cave's existence and its images were not made public until 1993. Using radiocarbon dating, archaeologists showed that the paintings had been executed in two batches, one 27,000 years before present (BP) and the other – including the great auks – around 19,000 years BP. Sea levels in the Mediterranean were lower then, and the entrance would have been accessible on foot. If the idea of great auks

The great auk images of Cosquer Cave, France. This photo is actually from an identical replica cave constructed nearby for visitors. (Courtesy Gabriel Beraha, Kléber Rossillon & Région Provence-Alpes-Côte d'Azur/3D MC)

frolicking in the Mediterranean seems unlikely, remember that further north, the ice sheets had pushed many species southwards. And, of course, as conditions warmed at the end of the Ice Age, some 11,000 years ago, the fauna (and flora) expanded northwards again.

It is generally agreed that it is pointless trying to interpret the 'meaning' of the images painted, scraped or engraved on the walls of caves and rock shelters by our ancient ancestors.[15] We can speculate, but no more than that, although many archaeologists have built careers from such speculations. What I find remarkable about the Cosquer great auks is just how well-observed they seem to be, especially compared with many illustrations executed in recent historical times (see Chapter 8). As far as I know, there's only a single speculation about the Cosquer great auk images, and it lies completely outside any of the other interpretive themes, such as art for art's sake or hunting magic. It is that whoever painted those three birds was documenting an ornithological extra-marital event, where one male great auk is attempting to mate with another male's partner. As indicated in Chapter 1, I have no doubt that great auks did behave in this way, but the idea that one of our ancient ancestors actually witnessed this behaviour may be a speculation too far.[16] What is certain is that those 'Upper Palaeolithic' Mediterranean cultures killed and ate great auks, since their remains have been found in many locations.[17]

Additional evidence that great auks were once abundant and occurred in areas other than those indicated by written records emerges from the wealth of bones collected by contemporary Dutch amateur archaeologists. As part of their land reclamation process, the Dutch government dredged huge quantities of sediment from the North Sea, dumping it along the seashore to create coastal defences. Private collectors soon found these sediments to be rich in fossils, and a community of citizen scientists emerged that continues to collect and document this material dredged from the sea. Analysis reveals dozens of great auk remains, mainly dating from the fourteenth century, but with some from as far back as 48,000 years ago. The birds are all thought to have died naturally, for none of the bones – mainly humeri (wing bones) – bear the cut marks that would have indicated that the birds had been deliberately killed and butchered.[18]

People first moved into Labrador around 9,000 years ago, and into Newfoundland 5,000 years ago, where they subsisted on marine mammals, fishes and seabirds. Great auk bones – together with those of other marine birds – have been found in kitchen middens all the way from northern Labrador to southern Newfoundland. It is generally assumed that apart from those settlements in eastern Newfoundland – nearest to the only known colonies on Funk Island and Bird Rocks, Magdalen Islands, Quebec (see map on p. 62) – these great auks were probably obtained at sea and were most probably birds that had dispersed from their colonies at the end of the breeding season.

Like many iconic animals from times past, the great auk was not merely a source of food. It was also – as exemplified by the Cosquer Cave artists – of symbolic significance. This is hardly surprising given its size, unique flightlessness and formidably powerful beak. At Cnip in the Outer Hebrides, some time between 400 and 150 BC, someone deliberately placed an entire great auk head within the walls of a house they were building – thought to be a Late Iron Age votive offering that said 'protect this house'.[19]

Great auks were held in high spiritual esteem elsewhere, too. At Port au Choix, on the west side of Newfoundland, there is a c.3,500-year-old cemetery (referrable to a group known as the Maritime Archaic people) in which some of the dead were interred with ceremonial objects derived from birds. One grave contained 200 great auk beaks (see page 44) that archaeologists think may have been the remains of a decorated feather blanket or cloak.[20]

A number of the graves at Port au Choix were also found to contain small, polished quartz pebbles – great auk gizzard stones. Some were associated with red ochre, a common and symbolic pigment used by past peoples. Birds of many different species – including other auks and penguins – deliberately ingest small stones to help grind up their food. In the great auk's case, this may have been to crush the bony skulls of fishes and the carapaces of crustaceans, or possibly as self-medication against gut parasites. It is easy to imagine our ancestors' awe at finding such 'gems' inside great auk gizzards. Similar stones have been found in the soil on Funk Island among the abundant skeletal remains of the great auks slaughtered there. Some of those collected by Owen Bryant in 1908 are now in the museum at Harvard in a little box accompanied by a note that reads 'from the gizzard of the great auk'.[21]

The Maritime Archaic people were not alone in cherishing gizzard stones. Independently, across many human cultures, the polished pebbles from the guts of many different birds were thought to possess magical qualities. The Greek physician Dioscorides, in the second or third century BC, reported that

Burial of an adult male at Port au Choix, Newfoundland, showing some of the 200 great auk beaks (black) and stone implements (grey). (Redrawn from Tuck, 1976)

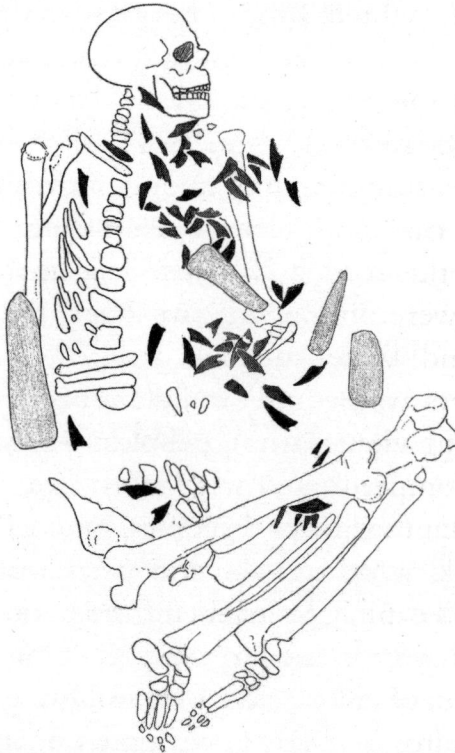

hanging the stone sometimes found in the stomach of a young barn swallow around the neck or the arm of a child would 'help the falling sickness' (epilepsy). This idea was still extant in seventeenth-century England. The Romans thought that the crystalline stone from a barnyard cockerel would confer invisibility, strength, courage and 'success with women'; the stone from a hoopoe 'laid upon the breast of a sleeping man, forces him to reveal any rogueries he might have committed'. By including them as grave goods, it is clear that the Maritime Archaic people at Port au Choix also considered great auk gizzard stones special, although in what way we will probably never know.[22]

The Port au Choix people were displaced and replaced by the Beothuk or 'Red Indians', who occupied Newfoundland and Labrador between AD 500 and 1829. So named not for the natural colour of their skin but for the red ochre with which they decorated their bodies, the Beothuk were the first indigenous people that Europeans encountered in their new-found land. Predictably, relations were not always amicable. The Beothuk lived on seabirds and fish in the summer and migrated inland during the winter to hunt caribou. Persecuted by European fishermen and settlers on the coast and by the Mi'kmaq tribes inland, the Beothuk were vulnerable to imported diseases like tuberculosis, and were soon in decline. Some of these colonial invaders were more 'enlightened' – if we can call them that – and viewed indigenous peoples as curiosities rather than competitors and adversaries. In 1828 the Scottish-Canadian explorer William Cormack came across a young Beothuk woman named Shawnawdithit living in Newfoundland's capital, St John's. Empathetic and intrigued, Cormack took her in so that he could learn about the Beothuk's language and culture. Although she had a good knowledge of birds, she was quite unaware of the great auk, which by then was extinct in North America. Shawnawdithit died at the age of just 30 in 1829 after contracting tuberculosis. She was the last of her people.[23]

It seems remarkable to me how different indigenous cultures exploiting birds for subsistence in distant parts of the world independently hit upon similar trapping techniques. The Ancient Egyptians hunting in the Nile marshes, for example, used tethered herons as decoys to lure wild birds within capture or killing range. In exactly the same way, the Beothuk used tethered geese and black guillemots as decoys.[24] Their sycamore or pine arrows, fletched with goose and bald eagle feathers, were tipped with a six-inch barbed lance made either from iron nails stolen from the Europeans or carved from caribou antler. Beothuk arrows have been found on Funk Island, suggesting that they may have used them to kill great auks, but they may also have caught great auks by hand, simply running them down.[25]

Once or twice each season, some Beothuk made the 60-kilometre (37-mile) trip in their high-prowed birch-bark canoes from Cape Freels out to Funk Island to collect birds and great auk eggs. They used the eggs to make a kind of sausage, mixing the dried yolks with seal fat and liver stuffed into sections of seal intestine. Funk Island lies so low it is invisible from the mainland, and the surrounding seas are often treacherous. It is a measure of the Beothuk's seamanship and navigational skills, together with the value they placed on the eggs and birds available there, that they were prepared to undertake this dangerous journey. I asked a knowledgeable sea kayaker about making such a trip and he thought it would take a good 10 hours to reach Funk Island, pointing out that the Beothuk would also have been confident enough to sleep occasionally at sea. Not surprisingly, some visits to Funk by the Beothuk coincided with those of European settlers. On 30 July 1792, when five Europeans were already present on the island collecting eggs, they saw two canoes of what they called 'savages' approaching. They fired at them, forcing them to paddle back to the mainland – a long, wasted trip.[26]

Great auk bones have been discovered at several archaeological Inuit sites in south-western Greenland, some dating from as early as 2,400 BC, but mainly from the fourteenth to nineteenth centuries AD. In the 1700s and 1800s, Danish colonial administrators, ministers and missionaries provide several first-hand accounts of the way great auks featured in Inuit folklore. It seems that the great auks killed by the Inuit in west Greenland at that time were not breeding locally, but were instead mainly young birds dispersing from colonies elsewhere, like Funk Island or Iceland. The birds were not common and were only ever encountered offshore among the islands and skerries between September and January, when most of them would have been in their distinctive immature or winter plumage.[27]

The missionary Hans Poulsen Egede, known as the 'Apostle of Greenland', arrived at what is now Nuup Kangerula on the west coast with his wife, Gertrude, their two children, and 40 'colonists' aboard the aptly named *Hope* in May 1721. Egede's hope was that there still existed Norse settlers in Greenland. He found only Inuit, whose lives he subsequently documented in sympathetic detail. An illustration from his account shows the way the Inuit hunted great auks from a kayak using a 'dart' attached to a bladder.[28]

A generation later, another Danish missionary, Otto Fabricius, provided the most comprehensive account of the way indigenous people of west Greenland hunted and used great auks. Living among the Inuit in the Frederikshab region between 1768 and 1773, Fabricius described in meticulous detail a single – dead – adult great auk he had examined. He also wrote an all-too-laconic account of what he considered was a recently fledged great auk chick.[29]

An image of everyday life in mid-eighteenth-century Greenland; the birds in the left foreground are probably great auks. (From Egede, 1741)

Because the Greenland great auks were seen only at sea, they were hunted – as Egede showed – using kayaks from which they were speared with a lightweight barbed dart designed to either pierce the bird's body or, as Fabricius says, 'to catch a wing, a leg, or the neck of the bird'. As several other accounts attest, great auks were superb divers. Fabricius tells us that the only way the Inuit could get them was for several kayakers to harry a single bird, shouting at it as soon as it surfaced, forcing it to dive repeatedly until it was exhausted and allowing them to approach close enough to dart it. Some were easier than others, a difference Fabricius attributed to age: naive first-year birds were easier than the smarter adults.

All parts of the bird were used or eaten, including the (cleaned) intestines and the fat, which was used in lamps. The black skin from the great auk's large, webbed feet was sewn together with sealskin to create small ornamental containers embroidered with coloured designs and attached to white sealskin bags. The oesophagus was used, too, tied off and inflated to create the bladder that was attached to the dart that enabled the Inuit to keep track of their victim as it dived to escape.[30]

The great auk's dense feathering was 'used for a kind of garment called a bird-skin parka which … men and women, especially during the winter, have under their other garments closest to their body with the feathers inward'. The Danish colonists also used great auk skins as bedclothes. Oblivious of the ongoing great auk plucking massacre on Funk Island, Fabricius also said that the great auk's 'wonderful firm and soft down could well be of good use in bedclothes if many of these birds were available'.[31]

In a kind of reversal of fortunes, the excavation of Hans Egede's house and kitchen midden in 1969–70, long after his death, revealed great auk bones from birds that he and his family had presumably eaten.[32]

People first settled in west Greenland around 2400 BC; prior to this (as early as 4000 BC) when the climate was milder, great auks were more widely distributed and probably bred in Greenland. As the Danish environmental historian Morten Meldgaard says, when the first Inuit first arrived, the great auks would have been 'unadjusted to human predation and it could be expected that especially the more accessible breeding colonies would have been depleted through over-exploitation by the rapidly expanding Saqqaq people. The hunting activities of the Saqqaq people [the pre-Inuit earliest inhabitants of southern Greenland] could thus have helped reduce the range of the great auk.'[33]

The last great auk seen in Greenland was killed by an Inuit hunter in a kayak during the winter of 1814–15 or 1815–16 off Qeqertarsuatsiaat (formerly Fiskenaesset). The bird's body was given to the director of the Royal Greenland Trading Company and, after passing through several more hands, ended up in Copenhagen's Natural History Museum of Denmark. Because the specimen is in winter plumage rather than breeding plumage (see Chapter 1) it was considered 'inferior' and was thus allowed to be handled by students.[34]

Indigenous peoples didn't worry too much whether birds' eggs were partly incubated when they ate them. Europeans, on the other hand, brought up on the eggs of farmyard fowl, wanted fresh, unincubated eggs without embryos. The Europeans' standard procedure on visiting a seabird colony was to 'tread down', that is, destroy, all existing eggs, in the knowledge that any eggs found on subsequent days would be fresh. It was a brutal and messy business. Many of the auk colonies in north-east North America were, like Funk Island, low-lying and the (fresh) eggs could simply be picked up from the ground.

There is no written evidence, but it seems likely that some of
the better-educated seamen – the officers or surgeons – visiting
Funk Island kept a few great auk eggs as souvenirs. They would
probably have selected clean – and hence fresh – eggs that were
easily emptied of their contents by cutting off the pointed end
of the egg. Cabinets of curiosities, or personal museums, were
increasingly common from the mid-1600s onwards, and eggs as
big and bold as those of the great auk would have made a fine
addition. Other than sometimes adding the bird's name
'Pingouin' to the eggshell, no one thought of noting the date or
place where their egg had been obtained.[35]

It took seventeenth-century European seafarers eight weeks
to cross the Atlantic, by which time, sick of their hardtack and
salted meat diet, they were desperate for fresh food. Accessible
seabird colonies in these new lands were like manna from
heaven. Describing how Funk Island's great auks were driven
'hundreds at a time' onto gangplanks into waiting boats,
Captain Richard Whitbourne said, in 1622, that 'it was as if
God has made the innocence of so poor a creature to become
such an admirable instrument for the sustenance of man'.[36]

One of Whitbourne's objectives was to promote settlement
in Newfoundland, so it is not surprising that he exaggerated
the ease with which supplies – including great auks – could
be acquired. The idea of New World plenitude –
incontrovertible evidence of God's benevolence – soon
became widespread. Knowing what we now know, it is easy to
see how this belief arose. Arriving in lands whose resources
had seemingly been barely touched by human presence, nature
must have appeared almost unbelievably profuse. From Jacques
Cartier onwards, seafarers alighting on Funk Island were
overawed by the sheer numbers of birds packed together and

the ease with which they could be killed. In much the same
way, the first settlers in Virginia celebrated the 'Fowles both of
the water and land, infinit [sic] store and variety', including,
of course, the passenger pigeon, whose abundance darkened
the skies for hours on end. When feeding on the ground, the
pigeons could – like the great auk – 'be knocked down by the
hundreds by a man with a short stick'.[37]

Other writers subsequently noted that Whitbourne's
description of great auks walking up planks into waiting boats
– like contrived filming of lemmings throwing themselves off

A contrived image, based on Richard Whitbourne's 1662 account, of
great auks walking up a narrow plank and throwing themselves into a
boat. (From Kearly, 1862)

cliffs into the sea in one of Walt Disney's films – had an apocryphal ring to it. Those who had been to Funk Island recognised that the 'slope of the rock and wash of the sea would render such a thing impossible'. Even so, the image of great auks literally boarding small boats, like animals entering Noah's Ark, was powerful, reinforcing the idea of plenitude in this new-found land and thus encouraging the settlement that would ensure sovereignty.[38]

Initially, the great auks on Funk Island were exploited for meat and eggs that were consumed by the seafarers themselves. From the 1600s onwards, as feather mattresses became increasingly popular in Europe, there arose an insatiable demand for feathers. On the island of St Kilda, some 40 miles off the west coast of Scotland, part of the tax levied on the locals by their rapacious landlords in the 1600s and 1700s was paid in feathers – mainly from the puffins on which the locals largely subsisted. On Funk Island, the much larger and more readily captured great auks provided an abundance of feathers.[39]

For more than 200 years, Funk Island's great auks were herded into stone pounds, boiled in large cauldrons and their feathers stripped off. With no wood on the island, the birds' fatty, featherless bodies provided the fuel for the fire beneath the cauldrons. Their charred and decayed bodies gradually formed the soil in which puffins now breed. It was these remains that formed the 'mountains of bones' discovered long after the bird was extinct. It also explains why John Milne, on visiting Funk in 1875 and finding no *broken* bones or damaged skulls, thought the birds had died 'peacefully'. Instead, it seems the birds were thrown into the cauldrons alive, as related by Aaron Thomas, an able seaman and privileged rating on HMS *Boston*, who visited Funk Island in 1794:

If you come for their [the great auk's] *feathers you do not give yourself the trouble of killing them, but lay hold of one and pluck*

*off the best of the feathers. You then turn the poor penguin adrift,
with his skin half naked and torn off, to perish at his leisure …
While you abide on this island you are in the constant practize
of horrid cruelties for you not only skin them alive, but you burn
them alive also to cook their bodies with.*[40]

Why, you might ask, would Funk Island's great auks persist in
returning year after year even as their friends and neighbours
were being slaughtered? There are two answers to this. First, great
auks, like their guillemot and razorbill cousins, were programmed
to return either to where they were reared or to where they bred
in previous years. Second, perhaps, for Funk's auks there was
nowhere else to go, or at least nowhere that was safer.[41]

Whereas indigenous peoples, and to some extent early
seafarers, exploited great auks in a way that had only limited or
local effects on the populations, it was the commercial
exploitation for feathers that sealed the great auk's fate on Funk.

As the English adventurer George Cartwright wrote in
1785:

*… it has been customary of late years for several crews of men to
live all summer on that island* [Funk], *for the sole purpose of
killing birds for the sake of their feathers, the destruction which
they have made is incredible. If a stop is not soon put to that
practice, the whole breed will be diminished to almost nothing,
particularly the penguins* [great auks]: *for this is now the only
island they have left to breed upon.*[42]

Even though the magistrates in St John's introduced a ban on
taking great auks and their eggs, enforced by public floggings,
it was too late. By the early 1800s Funk Island's great auks
were gone.[43]

The disappearance of the great auk from Funk Island was the result of two waves of greed: the first for meat, the second for feathers. Would those Europeans who slumbered on feather mattresses derived from the brutally slaughtered auks have slept so soundly had they known the price of their bed?

There was also a third greedy wave. It was a posthumous one, and inflicted no damage on the great auks.

Visiting Funk Island in 1841, the Norwegian naturalist Peter Stuvitz was amazed by the 'enormous heaps of bones' and by the stone enclosures into which the great auks had been driven before having their feathers stripped away. Knowledge of this skeletal abundance encouraged others to visit the island. In 1863, Thomas Molloy, the United States Consul to Newfoundland, was granted permission from the local government to mine [sic] the remains of great auks on Funk Island. He obtained no less than 35 tons of decomposed organic material, five tons of which were sold locally at $19 a ton. The rest of this 'guano' was shipped to settlements on the US east coast to fertilise the gardens of wealthy Americans.[44]

Thirty years later, when the British geologist and naturalist John Milne arrived at Funk Island, he reported that Molloy's 'guano' was 'like dark-coloured peat, [and] almost wholly composed of bones'. In just half an hour of digging he found the remains of 50 great auks. Because none of the beaks had their horny covering (technically, the ramphotheca), Milne assumed that the birds were interred 'at a remote period'. He was also amazed to find a plant rootlet growing through the neural canal of one bird 'so as to firmly fix the vertebrae in position'. Although as we've seen he assumed (incorrectly) that the birds had died peacefully, he added: 'Nevertheless, it may be that they were the remains of some great slaughter where the birds had been killed, parboiled, and despoiled only of their feathers; after which they were thrown in a heap, such as the one I have just described.'

It was Frederic Lucas in the 1880s who recognised the
scientific potential of Funk Island's great auk graveyard. The son
of a wealthy merchant seaman, Lucas received little in the way
of formal education, but instead travelled 'in the good days of
sailing ships' with his father on two long round-the-world trips,
the first at the age of nine, the other as a teenager. In between,
he collected birds' eggs and skins, insects and postage stamps,
and those trips inspired an interest in marine birds. He had an
uncle 'who was an amateur (and, I came to realise, a very
indifferent) taxidermist, and stuffing birds seemed like a very
pleasant and easy way of earning a livelihood'. In the 1800s the
only employment for anyone with a serious interest in birds
was at a museum. Lucas was lucky, for despite his limited
schooling and overt disdain for science, he had an outstandingly
successful career, publishing over 350 scientific papers.[45]

In 1882, after several years employed as a 'preparator' of
natural history specimens, he secured the position of Chief
Taxidermist at the National Museum in Washington DC. By
this time the fate of the great auk was well known, but Lucas
had also heard hints of the immense wealth of osteological
material still available on Funk Island. Desperate to see for
himself, in 1885 he asked his boss, Spencer Fullerton Baird, if
he could organise an expedition to Funk to obtain specimens.
Baird refused on the grounds of cost. Two years later, however,
Lucas, together with William Palmer, was allowed to join an
official mackerel-finding mission to Newfoundland waters
aboard the *Grampus*.

As Lucas recounted: 'A more harmonious party probably
never cruised together.' The mackerel were notable by their
absence, but the great auk bones were abundant.
Overwhelmed by what he was able to collect in just two
days on the island, Lucas wrote: 'It is no exaggeration to say
that millions of Garefowl [great auks] gave up their lives on
these few acres of barren rock.' And this was after Molloy

had removed his 35 tons of great auk 'guano' and bones. Lucas was struck by the differential survival of different bones, but was disappointed to find no intact skulls (they had invariably broken at the point where the upper mandible would have flexed upwards in life, at the front of the cranium), and only a single sternum and pelvis in good condition. On the other hand, he found a lot of humeri (upper forearm wing bones) – 1,400. Lucas was also disappointed not to find any egg membranes, as Milne had done. The 'inner linings' of eggs, as Milne referred to them, are remarkably thick in great auks, as they are in guillemots, and are clearly visible in some blown eggs in museums. In life, these robust membranes probably helped strengthen the egg and protect it from the rough and tumble of incubation on bare rock with no nest.[46]

Another point perceptively noted by Lucas was that 'the number of bones from young birds is extremely small, but this all but total lack of them is readily accounted for by the fact that after the merciless slaughter of the Auks had commenced, few, if any, eggs were allowed to hatch.' The fate of young great auks between hatching and leaving the colony has been a long-standing mystery, and one we'll return to in Chapter 6.[47]

Lucas was also disappointed not to find a great auk mummy, as Milne had done 24 years earlier. This is the only sure way of finding all (or at least most of) the bones of a particular bird and reconstructing an authentic skeleton. Of Milne's three mummies, two were sent to Professor Alexander Agassiz at Harvard and one to Alfred Newton in Cambridge. Newton carefully removed the dried skin to create an almost entire skeleton. This he sent to Richard Owen, at the time Britain's greatest anatomist. He, in turn, used the specimen to produce the first technical description of the great auk's skeleton. The mummy had lacked part of its left side, but Owen borrowed

the missing bones' equivalents from John Hancock, in
Newcastle, to complete the skeleton. Owen's careful
investigation of the great auk's skeletal anatomy enabled him
to compare it with other members of the auk family and with
the penguins, to try to infer who was related to whom in the
great tree of avian life. The similarity in body form and lifestyle,
to say nothing of their shared name, had suggested to some
that great auks and penguins were sister taxa with a common
ancestor. Owen showed that this was unequivocally not the
case: great auks and penguins are utterly distinct. The similarity
in both external and internal (that is, osteological) appearance,
however, confirmed the close relationship between the great
auk and the razorbill and guillemots.[48]

Funk Island was a killing field. As Frederic Lucas wrote,
'Countless myriads of this flightless fowl had been hunted to
the death with the murderous instincts and disregard for the
morrow so characteristic of the white race.' A mass of
slaughtered innocents; a grave despoiled by fortune seekers.

Great auk mummy from Funk Island. (Photographer unknown; image
courtesy of Errol Fuller)

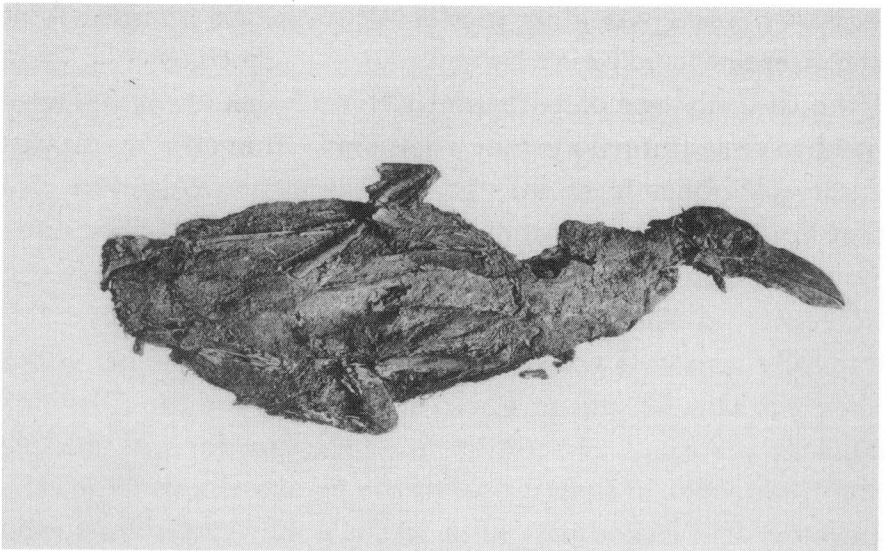

Their activities reminded me of the grubby opportunists who stole tons of mummified cats and birds from ancient Egyptian tombs or robbed human graves at Waterloo, in both cases to grind up and sell as fertiliser. Thirty-five tons of desiccated great auk remains ground into fertiliser – this is both an indication of the numbers of birds that once bred on Funk, and undeniable evidence of the extent of human greed.

Unintentionally, no doubt, the fertiliser brigade left sufficient remains for a different and more benign kind of body snatcher.[49] Funk was, and still is, a great auk graveyard. It is also a contradiction, with a million bird bodies underground and similar numbers alive on the island's surface. I visited Funk Island in 1980, arriving after a pleasant four-hour journey on calm seas under blue skies. The sheer number of birds, the intensity of the light and the overwhelming sense of pilgrimage evoked an enduring sense of euphoria. Half a million common guillemots, hundreds of gannets, Atlantic puffins, fulmars and other seabirds, gave me a vivid sense of what this tiny island packed with great auks must have been like prior to João Álvares Fagundes's arrival. My rapture was temporarily lessened, however, by the eight-hour vomit-splashed homeward journey in heavy seas. I was lucky; having declined the pickled rabbit and rum proffered by the crew, so avidly accepted by colleagues, I was the only one of my team that wasn't sick.

That visit, among all those birds, and being able to unearth great auk bones from the soil, still resonates with me today. But Funk Island wasn't the great auk's final resting place. That was Iceland.

Chapter 3

The Auk and the Walrus

There's an old joke in Iceland that asks: what is the difference between the walrus and the great auk? The answer is 'nothing': they were both negatively affected by the arrival of Norse invaders in the ninth century. Both were valuable as a source of protein, and the walrus's ivory was an especially treasured trade item. The two species were superbly adapted to life at sea but were far from agile on land, where they were hopelessly vulnerable to human hunters. As is evident from their skeletal remains in kitchen middens scattered around Iceland's rugged coast, both species were once abundant. By the eleventh or twelfth century the walrus was extinct in Iceland. The great auk, however, managed to cling on until the nineteenth century.[1]

The parallel effects suffered by these two species at the hands of human invaders make an important point. Everywhere humans have colonised there has been a concomitant loss of species by over-exploitation. We have already seen this among the great auks in Newfoundland and Greenland, and it was true in Iceland, too.

Across the entire North Atlantic, the great auk was only ever known with certainty to have bred at just *seven* locations. By the time the Funk Island colony was extirpated around 1800, the Icelandic population was the only one left, although no one knew that at the time.

Seven locations is a staggeringly small number, compared with the many hundreds of colonies of other auk species in the North Atlantic. Biologists and historians have wondered whether this ridiculously small number of breeding colonies was a natural feature of the great auk or a consequence of early human persecution before records were kept.[2]

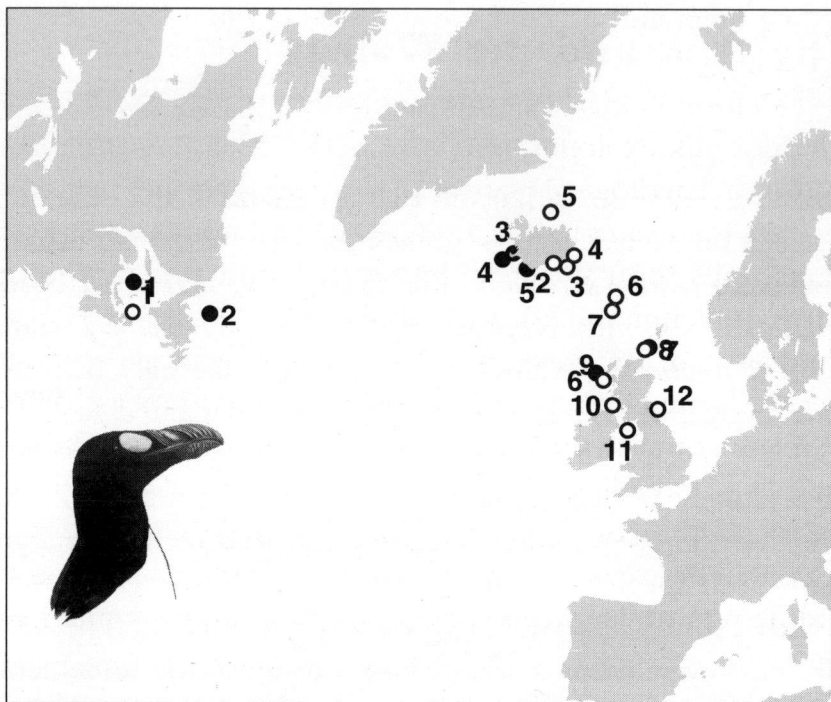

Known great auk colonies: Rocher aux Oiseaux (Bird Rocks), Magdalen Islands, Quebec, in the Gulf of St Lawrence (1), Funk Island, Newfoundland (2), Eldey, off Reykjanes Peninsula, Iceland (3), Geirfuglasker, off Reykjanes Peninsula, Iceland (4), Geirfuglasker, Vestmannaeyjar, Iceland (5), St Kilda, west of Scotland (6), Papa Westray, Orkney (7). From Nettleship & Evans 1985. For putative colonies see Appendix 1, p. 227.

Successive writers have suggested that the answer to this question, not surprisingly, can bes found in the great auk's inability to fly. The bird may have been naturally scarce because flightlessness limited the number of suitable breeding sites that lay within commuting distance to feeding areas. Extant auks – puffins, razorbills and guillemots – fly from their breeding colonies to their feeding grounds, often several times a day. The great auk's more limited foraging range would restrict the number of islands with suitable breeding habitat. As human persecution drove the birds to more and more remote locations, breeding sites with the right attributes – commutable feeding areas and access from the sea – would have become increasingly few.

The chronological link between the appearance of humans and great auk exploitation and extermination is written large in the archaeological record. On the island of Sanday in Orkney, for example, the number of great auk remains fell dramatically soon after people arrived (some time in the late fourth millennium BC), indicating that the birds had been exterminated. By contrast, at the same site in Orkney and over the same period, the numbers of gannets remained high right up to the sixteenth century, even though their limited number of breeding colonies and geographic distribution is similar to that of the great auk. The major, and obvious, difference between the two species is that the gannet can fly. And despite millennia of incredible persecution, gannets still survive.[3]

In addition to flightlessness, the great auk's demise may have been accelerated by a fascinating feature of their biology, referred to somewhat esoterically as the 'Allee effect'. Swimming against the tide of conventional 1930s ecological thought, the pioneering American ecologist Warder Clyde Allee suggested that certain animal species do better at higher rather than lower population densities. Most ecologists took it for granted that as an animal's population increases, competition for food or breeding sites also increases, and individuals survive and reproduce less well. Allee turned this on its head under the ungainly title of 'positive density-dependence'. He recognised that for certain species, when their population is small, they lack the assistance of others – in finding food, for example, or defending themselves from predators – precipitating an exponential population decline.[4]

When I started studying guillemots on Skomer Island, Wales, in 1972, the Allee effect was all too apparent. Fifty years previously the breeding guillemot population there had been huge, probably 100,000 birds. By the early 1970s that population had decreased by 95 per cent as a result of oil pollution during and after the Second World War. This species depends on breeding at high density to protect itself from the depredations of gulls and ravens, and it was clear that guillemots at much lower numbers were extremely vulnerable. Exactly the same may have been true of great auks, accelerating their decline towards extinction.[5]

Iceland was the great auk's last hold-out, but compared with Funk Island there is less historical information available. Had they or their eggs been important, as a source of food in the thirteenth and fourteenth centuries we might have found great auks featuring routinely in the Icelandic Sagas. In fact, the Sagas contain only a single mention, where the great auk is just one of a list of 120 bird species, suggesting that by the date this was written (c.1220 – some 300 years after Iceland was first settled) people had destroyed the great auks at all easily accessible locations. Excavations of kitchen middens dating from the centuries before the Sagas, however, show that great auks were abundant when the first settlers arrived.[6]

Great auks persisted at a handful of locations in Iceland because, unlike Funk, these islands were not on a major maritime trade route. An account from 1589 states that the great auk was then abundant ('in plenty') on several islands off Iceland. One of these islands must have been Geirfuglasker – literally Great Auk Island – the penultimate in a series of skerries collectively known as Fuglasker (bird skerries) off the Reykjanes Peninsula in south-west Iceland. A remarkable manuscript from 1770 includes a drawing of boats anchored offshore and three men killing great auks on the skerry. Getting to this particular island was a risky venture. In 1628, 12 men drowned there, and in 1639 two boats were lost attempting to land. Even so, after looking at old records and interviewing

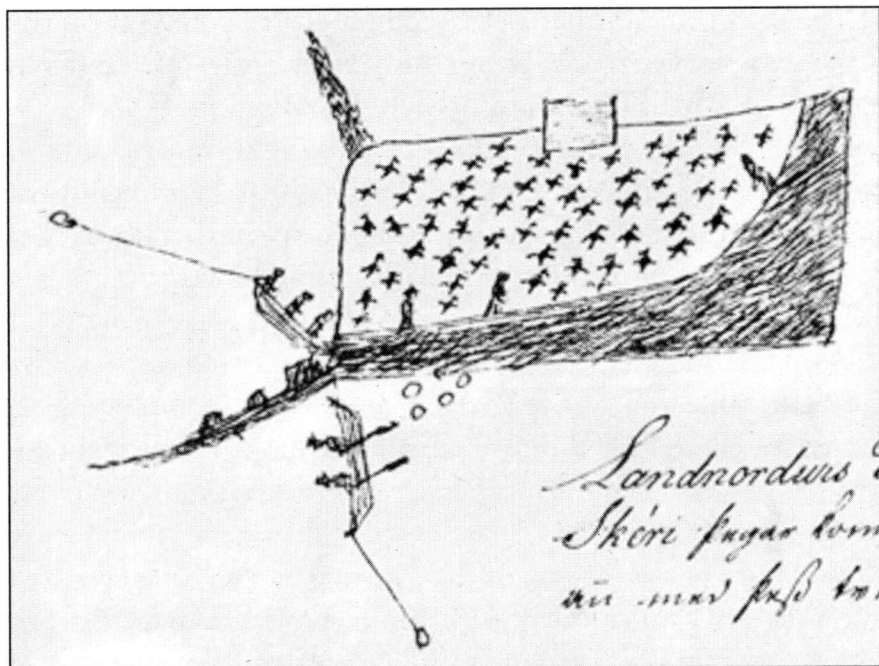

Geirfuglasker, from Sigurðsson 1770. I like this drawing, not for what it represents, but because of its naive execution and the fact that it might equally well have been painted or scratched onto a cave wall thousands of years earlier. In contrast to the imagined images showing the destruction of great auks at Funk Island (see figure on p. 52), the sheer simplicity of Sigurðsson's image has an immediacy that screams authenticity. (From Bardason, 1986)

some old men in 1858, the ornithologist Alfred Newton came to the conclusion that from about the middle of the 1700s, Geirfuglasker was 'constantly visited by fowling expeditions' filling their boats with great auks and eggs.[7]

In his account of Geirfuglasker, Guðni Sigurðsson (1770) discusses the dangers of visiting the island: 'to go there is to place life and death on an even chance'. On the other hand, a successful visit was immensely rewarding in terms of birds and eggs. As though to emphasise the risks, he says: 'in the year 1732, after a lapse of seventy-five years, the skerry was visited and two huts and some withered human bones were found'.

He adds that on another occasion, three men stranded on the island had survived by 'eating sun-dried birds and drinking rotten eggs for half a month', before being rescued.[8]

In late July 1808, Geirfuglasker was visited by the crew of HMS *Salamine*, commanded by John Gilpin, 'where they spent the day killing birds [unspecified seabird species] and treading down their eggs and young'. Similarly, in July 1813 the schooner *Faeroë* becalmed off the skerries, and finding a great many great auks on Geirfuglasker, 'killed as many as they could, and loading the boat quite full, yet left many dead ones … intending to return for them'. They didn't return, because the wind got up and they had to leave. On arriving at Reykjavik, they had on board no fewer than 24 great auks and 'others which were already salted down'. One of the 24 corpses was presented as a gift to the Bishop of Reykjavik. He passed it on to an English friend, and that bird, suitably stuffed, eventually became part of the Foljambe family's fine collection of mainly British birds at Osberton Hall, not far from where I live in Sheffield and where I was fortunate to be able to see it in July 2023.[9]

The only other documented visit to Geirfuglasker was in late June 1821 when Frederik Faber, a Danish merchant, who we will meet later, along with Count Raben and his servant, failed to find any great auks there.[10]

In early May 1830, a protracted submarine volcanic eruption caused Gierfuglasker to disappear beneath the waves. Great auks had lingered here longer than anywhere else precisely because the skerry had been both distant and unpredictably accessible. With Gierfuglasker gone, the remaining birds shifted to breed on the aptly named 'Fire Island' of Eldey. Part of the same chain of bird skerries, Eldey is closer to the mainland, instantly rendering the birds more vulnerable than previously

The Reykjanes Peninsula in 1780, showing Geirfuglasker and Eldey.
(Courtesy of islandskort.is)

to human depredation. Eldey was also known as 'Meel-saeken' (the Meal Sack) by Danish sailors because 'its appearance is grotesquely like that of a monstrous half-filled bag of flour', a resemblance enhanced by its dusting of white, flour-like bird droppings.[11]

That Geirfuglasker's great auks had moved to Eldey was discovered by Brandur Guðmundsson in 1830, possibly while hunting other seabirds. In that year, he made two trips to Eldey, killing 12 or 13 great auks on the first, and a further eight on his second visit. He sold their bodies to local merchants living at Keflavik and they sold them on to collectors elsewhere. The discovery of Eldey's great auks sparked a scramble for specimens, fuelled by the news of the species' scarcity and the desperation of museums to obtain specimens before it was too late.[12]

In the early 1800s ornithology was finding its feet. As it did so, the need, and specifically the desire, for specimens in the form of skins, skeletons and eggs was insatiable. Without specimens there was no ornithology. Since there were no readily available binoculars to watch birds, almost the entire focus was the description and classification of birds, and for that one needed reference material. There were few national museums and most of the people interested in birds were wealthy and obsessive men. Ornithology was simultaneously a quest for both knowledge and status: more specimens of more genera from more regions of the world. Better guns, better education and better trade and travel routes all facilitated collecting, and the extent of private collections was extraordinary; thousands of men and a handful of women collected birds' skins and eggs.

Even though no ornithologist ever saw a live great auk, the bird's existence was well known since it featured in most of the bird books of the time. These included the Danish scholar Ole Worm's Museum catalogue *Museum Wormianum*, published posthumously in 1654, in which he describes keeping a great auk (a gift from the Faroes) as a pet. John Ray and Francis Willughby in their encyclopaedia of ornithology of 1676 provide a brief description of the great auk based on a dried skin (probably from Newfoundland) in the Royal Society's 'Repository'. They illustrated their account using Ole Worm's image of the great auk – at one time said to be the only one drawn from life.[13]

The French public may also have been familiar with the live bird after seeing the one that Louis XIV kept alive at the palace of Versailles. This too may have been drawn 'from life', albeit rather quaintly, swimming on a pond rather than the sea.[14]

Perhaps the best known of all great auk accounts was Martin Martin's, based on his interviews with the inhabitants

Ole Worm's captive great auk from 1654. The white ring around the bird's neck was a collar by which Worm controlled the bird, and not part of the bird's plumage, as some subsequent illustrators assumed. (Biodiversity Heritage Library)

of St Kilda when he was there in 1697. Martin did not see one himself, and indeed the great auk may have ceased breeding on St Kilda several decades before his visit. By the early 1700s, the great auk was no more than an irregular visitor to St Kilda, with one captured in July 1840 (the actual year is uncertain). Five men had sailed from the main island of Hirta to the most remote part of the archipelago, Stac an Armin, where they came across a sleeping great auk. The bird was captured, its feet were tied together and it was taken to the men's bothy (small hut). For three days the bird bellowed in protest, making 'a great noise' and opening its mouth when anyone approached. When a storm blew up, the rowdy and unfamiliar bird was deemed to be the cause. Assuming that it was a witch, the men stoned it to death, dumping its body behind the bothy. Belief in witches persisted well into the nineteenth century, especially, as in this case, among 'island men, reared in lonely isolation and brought up in a world steeped in superstitions'.[15] Martin's account has been widely quoted and revered, but it was neither as 'first hand' nor as reliable as was once thought.[16]

Carl Linnaeus included the great auk in his great taxonomic work of 1758, *Systema Naturae*, but he, like all subsequent ornithologists, never saw one alive. Indeed, in contrast to Willughby and Ray a century earlier, there is no evidence that Linnaeus even saw a skin, for at that time they were very rare. Instead, he probably relied on the accounts of a handful of trusted ornithologists.[17]

As the hankering for great auk specimens increased, so did the number of natural history dealers. Gone from Newfoundland, the search for great auk specimens now focused entirely on Iceland, which at this time was under the Danish crown. Due to conflict between Denmark and Britain as part of the Napoleonic wars, the Icelandic trade in great auks flowed almost exclusively through Copenhagen. Specimens were scarce, and they were expensive. As we saw earlier, George Savile Foljambe managed to get one via the Bishop of Reykjavik in 1813. The German natural historian Johann Casimir Benicken also secured a skin through a contact in the Royal Greenlandic Trading Company in 1821. By contrast, Frederik Faber, who spent more than two years looking for great auks in Iceland between 1819 and 1821, failed to obtain a specimen. Recognising that there was money to be made from great auks, Danish dealers commissioned local Icelanders to take the dangerous journey out to Geirfuglasker.[18]

There were several prominent dealers, including Carl F. Siemsen, Georg W. Brandt, an apothecary named Möller and another known only as 'Israel of Copenhagen', all living in or working out of Reykjavik. After Geirfuglasker disappeared beneath the waves in March 1830, the dealers' attention switched to Eldey Island. The removal of great auks from there was like emptying a bath of water.

Here's the toll:

Geirfuglasker
1808–10: 2 [but others killed by crew of *Salamine* in 1808?]
1813: *c*.20
1814: 7
1815–22: 0
1823: 2
1824–27: 0
1828: 1
1829: 2

Eldey
1830: 20
1831: 24
1832: 3
1833: *c*.13
1834: 9
1835–39: 0 despite regular expeditions
1840: ?
1841: 4
1842: 0
1843: 0
1844: 2

Not all of these great auks were transformed into museum skins. Some were plucked and eaten. But after 1828, as the dealers began to realise their monetary value, great auk skins were all sent away to Copenhagen, Hamburg, Flensburg and a few to England. The local Icelanders, however, did eat the skinned bodies.

The annual 'bag' of Icelandic great auk specimens listed
above comes partly from Alfred Newton, and partly from the
English-born German William Preyer, who conducted
interviews with the dealers. His data formed part of his PhD
on the great auk, completed in 1862 when he was just 21.
Alfred Newton, who considered himself *the* great auk expert,
was dismissive of Preyer's efforts. The origin of Preyer's interest
in the great auk is a mystery, and after his doctorate he switched
from natural sciences to medicine, ending up as a pioneering
child psychologist. He was inspired by Darwin, and the two
corresponded in the 1870s and 1880s.[19]

In 1844, the dealer Carl Franz Siemsen, who acted as an
agent for various foreign scientists and museums, received
an order for a great auk. He offered local farmers 300 Icelandic
krona for one, dead or alive, and in early June that year 14
men rowed out to Eldey with the express purpose of acquiring
a great auk. Three of the men made the dangerous landing
and ascent, and on seeing *two* great auks among the guillemots
and razorbills, gave chase. The birds 'showed not the slightest
disposition to repel the invaders'. The two great auks simply
'ran along under the high cliff, their heads erect, their little
wings somewhat extended', uttering 'no cry of alarm, and
moved with their short steps, about as quickly as a man could
walk'. One of the men, Jon Brandsson, drove one bird into a
corner 'where he soon had it fast'; the other two men chased
and caught the second bird. Both birds 'were strangled and
cast into the boat'. A single egg was seen, but on being found
to be damaged was left. Getting back into the boat in what
they called 'Satan's weather' was tricky, and Jon had to jump
into the sea clutching a rope, after which he was pulled into
the boat.[20]

Carl Franz Siemsen, the German Reykjavik-based great auk dealer responsible for the deaths of at least 21 great auks, including the last two birds. (Courtesy National Museum of Iceland)

Saying that the two birds were 'strangled' implies that they were killed by depriving them of breath. Auks are built very robustly and have a considerable capacity to hold their breath. Suffocating them would have taken too long, especially as the men were keen to get away from the island because of the weather. I suspect that the birds' necks were broken.

Subsequent expeditions to Eldey found no further great auks. Although they did not know it at the time, the birds Jon Brandsson and his two colleagues killed were among the very last great auks.

In 1984, the Swedish ornithologist Sven-Axel Bengtson published a particularly comprehensive assessment of the great auk's ecology. His essay was commissioned by the editor of *The Auk* to celebrate the centenary of that American journal of ornithology, whose eponymous cover depicted a great auk. Bengtson wrote: 'In my opinion, the decline of the great auk commenced long before man is known to have caused havoc

in breeding colonies in the mid–16th century and onwards.'
Using information from Icelandic sagas to infer climatic
changes, he explored the possibility that environmental issues
had reduced the abundance of great auks to the point of no
return. 'There is good evidence that a climatically severe
period preceded and coincided with the period when man
was dogging the great auk.' As support for his idea of
environmental change contributing to the decline, he referred
to what were then, in the 1980s, the recent declines in the
abundance of fish on which Norwegian puffins depended.
The result – as we now know – was that puffins and other
seabirds had to fly further in their search for food, ultimately
paying the price, as witnessed by their dramatic population
declines. Bengtson argued that the great auk's flightlessness
and limited foraging range made them especially susceptible
to such environmental changes.[21]

In a cautious conclusion, Bengtson wrote: 'In the absence of
more detailed information about the rate of decline and bird
populations, hunting pressure, and environmental changes, we
cannot separate the effect of climate.' When he wrote those
words in 1984, he was referring to non-anthropogenic climate
change. The idea that we were already experiencing global
climate change seemed utterly remote, but the first signs were
already apparent.

Whether or not climate change *was* involved, the impact of
humans on great auks was all too obvious. Earlier I described
as an arms race the conflict of increasing human technology
versus the ability of the bird to find ever more remote locations
to breed. It was a battle lost by other flightless birds, too, and
for the same reason. Following the Polynesian people's arrival
in New Zealand around 1300, moas were exterminated by
the late fifteenth century. Madagascar's enormous elephant
birds seem to have coexisted with humans for rather longer,
before being wiped out around AD 1000. Dutch seafarers

discovered the dodo on Mauritius in 1598 and by 1681 it too was extinct, with the closely related Rodrigues solitaire gone by 1730.[22]

The great auk is an exemplar of the impact that people have on different species across the world. It is a constant, repeating and now entirely predictable scenario: human invasion, exploitation and extinction. It is a pattern that will probably see our own extinction. The great auk's demise may seem like a mere curiosity to some, but it is a lesson we ignore at our peril.[23]

As we are now all too aware, the two auks killed in 1844 were among the very last of their kind. In the 1850s, though, a flame flickered in the hearts of a handful of ornithologists who hoped that a few birds might still survive. Some of them went to look.

Chapter 4

Three Men in a Boat

Being called a 'wally' is a gentle insult. It means foolish or naive, in much the same way as the Spanish name *Pájaro Bobo* was used for the 'silly bird': the great auk. John Wolley, the nineteenth-century great auk enthusiast, however, was anything but foolish.

Not well-known in the annals of ornithology, Wolley was nevertheless among the smartest of his generation. Had he survived longer than a mere 36 years, his would be a name indelibly and unforgettably linked with the study and conservation of birds, and especially the great auk.

Born in Matlock in Derbyshire's Peak District in 1823, John Wolley was from a privileged and educated family. His father, a clergyman, accumulated and then donated a valuable collection of ancient manuscripts to the British Museum. In 1836, at the age of 13, Wolley was sent to Eton, where he met fellow pupil George Dawson Rowley, who, like Wolley, had 'more than the usual taste for birds'-nesting'. In 1842, Wolley went up to Trinity College, Cambridge, where he spent more time collecting insects than studying; perhaps inevitably, he left without completing his undergraduate degree. In 1845, he travelled through southern Spain to Tangiers where he 'unexpectedly found domiciled a keen egg-collector, at that time known to few naturalists in Europe and perhaps to none in England', the ardent oologist François Favier. As Wolley's colleague Alfred Newton later wrote: 'The discovery of M. Favier and the treasures he possessed may be said to have been the turning point of Wolley's life, with entomology being overtaken by oology.'[1]

Despite his less than outstanding performance at Trinity, Wolley was clearly academically able. After his return from

Spain he became fascinated by the dodo. Long extinct, the dodo was poorly known, and Wolley began to compile everything he could find about the bird. Then, discovering that a more senior ornithologist, Hugh Strickland, was on a similar quest, Wolley gave him all his notes – the kind of selfless gesture that one might struggle to witness today, and a clear reflection of Wolley's generous temperament.[2]

In December 1846, Wolley bought a great auk egg for 28 shillings from the Reverend D. Barclay Bevan. This extraordinarily beautiful egg – cream with red markings – from the great auk's last known breeding site on Eldey represented the foundation of what was to become an obsession. Wolley's interest was further piqued when on a visit to Iceland in 1845 an old school friend, William Milner, had

John Wolley. The photograph was taken in the winter of 1858–59, about a year before he died. (Photo by Edward Joseph Lowe, from Volume 1 of *Ootheca Wolleyana*, 1864)

failed to secure a specimen of the great auk, leading to 'suspicions as to the bird's fate'.[3]

In 1847, at the age of 24, Wolley embarked on a three-year medical course in Edinburgh. He continued to form 'a collection of birds eggs, all the specimens of which should be thoroughly trustworthy, and by consequence not only serviceable to but worthy of a scientific study'. The issue of trustworthiness with regard to biological specimens like eggs was captured in the following:

Then there was the consideration of the culpable carelessness as to verification of specimens displayed by so many of the owners of even large collections, and the futile arguments by which they strove to persuade themselves that this or that egg, bought from a dealer who had a plausible story to tell, was indeed the treasure it was asserted to be.[4]

Written by his dear friend Alfred Newton after Wolley's death, this sounds suspiciously like a snide reference to the untrustworthy dealers that Newton was all too familiar with.

The summer of 1849 saw Wolley in the Faroes collecting the eggs and skins of various birds, but also making enquiries about the great auk. It was not known for certain whether the great auk had ever bred on the Faroes, even though some locals were familiar with it. Wolley wrote, 'An old man … had seen one fifty years ago [i.e. around 1800], sitting among the *Heldafuglar*, that is young guillemots and other birds in the low rocks, and an old man told him it was very rare'.[5] It is remarkable that Wolley's informant refers to the 'young guillemots' on the low rocks. My long-term study of guillemots, based on colour-ringed birds of known age, shows that congregating on the low tidal rocks beneath a colony is exactly what three-, four- and five-year-old guillemots do before they start to breed at age six or seven. It is also what I'm sure young great auks would have done, reinforcing my suspicion that the great auk seen around 1800 was an immature bird. Prior to this, great auks probably occurred more frequently on the Faroes, for an adult was captured in the 1650s, and the local priest, the extraordinary Lucas Debes, sent it to his mentor Ole Worm, in Denmark, who kept it as a pet for several months.

During his time on the Faroes, Wolley met Daniel Joënsen, who had been captain of the 12-gun schooner *Faroë*, which had sailed to Iceland in 1813 to secure provisions for 'the half-starved' Faroese. The Court of Copenhagen had prohibited, 'on pain of death, all intercourse with the British' because of an ongoing dispute. As a result, the 'unfortunate Faroese were nearly reduced to a state of starvation'. In an act of desperation, the governor sent the *Faroë* to Iceland for much-needed supplies. On reaching Cape Reykjanes off the south-western corner of Iceland, however, the ship was becalmed. The crew were able to land on one of the seabird skerries where – as we have seen – they killed at least 24 great auks.[6]

Learning all this only increased Wolley's fascination with the bird.

After several years of corresponding about their shared ornithological interests, Wolley and Alfred Newton finally met in 1851. Six years younger than Wolley, Newton was also from a wealthy family, with the Newtons' wealth ultimately derived from Caribbean sugar plantations. As a boy, Newton had been fascinated by birds, shooting them and catching them to keep in cages. He went up to Magdalene College,

Cambridge, as an undergraduate in 1848, graduating with a BA in 1853. Apart from a few trips, Newton never left Magdalene. He and Wolley remained friends, and it is probably fair to say that John Wolley's short life changed Newton's in the most profound way.

In 1853, on yet another trip, Wolley travelled in northern Scandinavia, where he lived with the Sami people, collecting more eggs and yet more specimens. On his way home, he stopped off in Oslo to meet Professor Japetus Steenstrup, who told him about an article on the great auk he had recently completed. Steenstrup's trailblazing study of the geographic distribution of the great auk was based on bones recovered from the kitchen middens across the entire sweep of the North Atlantic. His account provided an intriguing and comprehensive picture of the bird's widespread occurrence in

Alfred Newton. I have tried without success to find a date for this photograph, but it must have been taken about the time he and Wolley went to Iceland. (Courtesy Errol Fuller)

earlier times. Spellbound, Wolley wondered whether there might still be some great auks alive. If there are, Steenstrup told him, they must be in Iceland.

There was more. The following year Wolley learned of an article boldly proclaiming that great auks *were* still to be found in Iceland. Convinced that the great auk was *not* extinct, he and Newton decided to see for themselves.[7]

On 21 April 1858, after much planning and preparation, they set off from Leith, Edinburgh's port, bound for Reykjavik on board the steamer *Victor Emmanuel*. Following their overnight stop in Tórshavn, the Faroese capital, Wolley and Newton could hardly contain their excitement at the prospect of their ship passing within sight of Eldey, the great auk's last known breeding site. With telescopes and the naked eye they scanned the waves, hoping against hope that they might spot the bird they had travelled so far to see. They didn't, but neither did they give up hope. Indeed, the main objective of their expedition was to get to Eldey to check for themselves.

Their host in Iceland was Vilhjálmur Hákonarson, a local leader and the foreman of the fishing boats they planned to use to reach their ornithological El Dorado. Hákonarson had made the trip to Eldey several times, and, on his first visit there in 1831, had killed no fewer than 20 great auks.

Wolley carried with him a stack of cash − £120 (equivalent to £10,000/$12,600 today), making both men fearful of being robbed. The money was to pay the crew of two ten-oared boats that would carry them to Eldey. There were to be 16 oarsmen in total, sufficient to allow some to rest during the 12-hour trip. With Wolley, Newton and their interpreter Geir

Zoëga, this made 35 men in total. Also in the two boats was enough food for a week in case they became stranded. The crew comprised farmers and fishermen, several of whom had made the trip before; it included Brandur Guðmundsson, who had previously killed 40 great auks.[8]

Prior to setting off for Eldey, Wolley and Newton spent three weeks in Reykjavik – a town of around 1,500 inhabitants, most of whom lived in poverty – talking to the small number of educated people, visiting the library and trying to obtain as much information as they could about the great auk. It was a productive time in as much as they found previously unknown accounts of the birds, but the weather was miserably cold and wet. Their research was hampered by their being 'in a chronic state of intoxication', thanks to the hospitality of their Icelandic hosts. On one occasion, they were taken aback to be served a

The island of Eldey gleaming tantalisingly on the horizon. (From Newton's *Ootheca Wolleyana*, Volume I)

meal of turnstones and purple sandpipers, which they identified after asking for the birds' heads to be brought from the kitchen. Finally, they set off from the capital on 19 May for Kirkjuvogor on the Reykjanes Peninsula. Two days on horseback brought them to Hákonarson's 'lofty farmhouse' that was to be their base until they boarded the boats. But the weather remained poor and quite unsuitable for a trip. One morning, however, the seas were calm and, in a state of great optimism, they prepared themselves, only for Hákonarson to call it off at the last moment. His caution proved prescient, for the seas were soon rough again. The weather didn't let up, and as the days dragged on, Wolley and Newton began to wonder if they would ever get to Eldey.

A change of tack was necessary. The two men switched to grubbing around in nearby middens in search of great auk bones, and talking to those who had previously taken part in the hunting trips. Because Hákonarson was popular and well respected, the locals were happy for Wolley and Newton – via Zoëga – to interview them, and to recount what they could remember. Some of those they spoke to had killed great auks, not only on Eldey, but also on Geirfuglasker before it disappeared in 1830. The interviewees were amused that Zoëga's Christian name was Geir, as in Geirfugl – garefowl, or great auk – joking that he had obviously been born for the job. It was he who asked the questions and interpreted the answers, while Wolley tried desperately to understand and write it all down in his notebooks.

How were the great auks hunted? How did they behave when chased? How did they run? How did they swim? What did they sound like? Who took part in that trip in 1844 when the last great auks were killed? Tell us now, in detail, about that last trip!

The dead birds were given to local Icelandic women to skin and stuff; the bodies boiled and eaten, and the crudely preserved skins packed up and sent on their way (see Chapter 3). Eggs were less valuable and, despite their tough shell, were vulnerable to being broken in the business of chasing the adults and boarding the boats in boisterous seas. However, eggs were taken when encountered, and earlier, while in Reykjavik, Wolley had found an account from 1760 of a great auk hunting trip to Geirfuglasker that said that the Danes would give 8 to 10 fish for a blown egg. Newton tried, but failed, to discover the value of 'fish'.[9] Any eggs the hunters brought back were emptied (blown), probably by the same women who skinned the birds, by making a hole with a big needle at each end of the egg and flushing out its contents, whether yolk and albumen or a developing embryo. This was a crude way of emptying an egg; oologists elsewhere later emptied eggs through a single hole on the side of the egg made with a special egg drill. Two holes at the eggs' poles weakened the shell, and explains why many of the great auk eggs now in museums are damaged, especially at their pointed ends.[10]

Wolley and Newton never did make it to Eldey; the bad weather never let up. They returned to England disappointed, but with a lot of new information. Soon afterwards, Wolley wrote to Newton announcing his engagement to the 22-year-old Jeanette Lorraine, hinting heavily that marriage would curb his and Newton's collaborative efforts. At around the same time Wolley, who had previously enjoyed good health, began to suffer from a 'painful languor' and lack of energy. Soon he was experiencing memory loss as well. On consulting a doctor in London he was told that he had 'an affection of the brain', with little hope of recovery. As Newton subsequently wrote, 'these fatal words were fulfilled to the letter; not many days passed before Wolley experienced another violent attack. After lapsing into unconsciousness, he died, without suffering, on 20 November 1859.'[11]

Wolley's death – most likely from a brain tumour – was a devastating loss both for his fiancée and for Newton, for whom Wolley was not merely a close friend – perhaps his closest – but also his mentor and a great source of inspiration. Wolley was just 36; his tremendous potential as an ornithologist was never fully realised.

People can be clever in different ways. Wolley and Newton were both talented scholars, but my impression is that Wolley was more intellectually agile, more innovative and more daring than Newton. Picking up his legacy, Newton benefited by following up on several of Wolley's ideas. There's no suggestion that Newton 'commandeered' these ideas; his aim was simply to honour his esteemed friend: 'It has been, and always will be, a matter of regret, that his [Wolley's] active mode of life and his premature death prevented his giving to the world the connected account of his discoveries.' Although Newton was later credited with these achievements, it was Wolley who first proposed a census of the birds of Great Britain; Wolley who suggested the need for a Seabird Protection Act; and Wolley who pushed to document the history of the great auk.[12]

To commemorate his friend, Newton decided to produce a catalogue of Wolley's egg collection. Somewhat pretentiously, perhaps, he named it *Ootheca Wolleyana*; the word 'ootheca' referred to the egg case of an insect, so essentially a collection of eggs, in this case owned by Wolley. Compiling it took Newton a long time, with the first volume published in 1864 and the last in 1905, for he was the ultimate procrastinator. Initially, Newton employed John Balcomb (of whom little is known) to paint the eggs for Wolley's catalogue, but Balcomb must have moved on because it was Henrik Grønvold – one the most accomplished bird artists of the day – who produced the superb great auk egg images for Newton and Wolley's *Ootheca*.[13] As well as his large collection of eggs and skins, Wolley also left all his great auk notes to Newton.

Wolley's original idea – hatched in the 1850s – to write a book on the great auk almost expired with him. It was saved by Newton, who continued to add everything he could find on the bird. At some point he enlisted Wolley's old school friend George Dawson Rowley to help produce the long-anticipated monograph. Dawson started to compile all the known information about the bird, tracking down the location of every egg and carcass. The result was two huge leather-bound volumes weighing nearly 7kg (15lb) each, entitled *Alciana*, Volumes I & II, and still owned by the Rowley family.[14]

Another scion of a wealthy Victorian family, Rowley was able to acquire two mounted skins and six great auk eggs. His and Newton's plan was that their great auk monograph should include a catalogue describing the provenance and current whereabouts of every great auk skin, skeleton and egg known.

The idea wasn't new. Stimulated by an article Newton published in 1860 celebrating Wolley's short life, a certain Alfred Roberts of Scarborough in east Yorkshire attempted to list all known great auk skins and eggs in a short piece in *The Zoologist* of 1861. Listing the location of 26 skins and 21 eggs, he concluded by sensibly saying 'Should any omissions or errors be detected I should be glad if any of your correspondents would favour me with the communications.' There were two immediate responses: one sensible – on skins – and one irrational – on eggs. Mr Robert Champley, also of Scarborough (did he and Roberts not talk to each other?), identified several skins that Roberts had missed, bringing the total number, he said, to 44. Then there was J. P. Wilmot's bad-tempered communication, pointing out contemptuously that one of the eggs listed by Roberts was a fake. A second somewhat boastful point was that he himself had once had three great auk eggs. One of these he had given to Beebee Bowman Labrey (mentioned earlier – see p. 14). Another he gave to his late friend, John Wolley, and one he kept for himself. The egg he had retained was illustrated in Hewitson's 1831 book *British Oology*, and was, of course, the most beautiful great auk egg Wilmot had ever seen.[15]

Incensed at being overlooked, perhaps, Wilmot said: 'How Mr Roberts could write this article without referring to Mr Hewitson's work [*British Oology*] quite surpasses my comprehension,' and 'How could the editor and publisher of *The Zoologist* allow Mr Roberts's misstatement to pass without comment?' The editor, Edward Newman, responded, exonerating the publisher. As for himself, he did not feel he had committed any error, and repeated Roberts' plea by inviting any 'gentlemen who possess or know of eggs … to record the fact at once in these pages: I may thus be enabled to make out a tolerably complete list'.

Roberts in turn responded, apologising for omitting Wilmot from his list, saying once again that he had deliberately asked readers to point out any omissions. As if 'sufficient proof of the utility of my remarks' was needed, he recounts that he received numerous letters 'from different parts of the Continent', adding to his stock of information. He also says that there was 'not the slightest reason' for Wilmot to pen his note in the way he did. The editor agreed and asked that any further letters 'stick to simple facts and not to taste or opinion'.

Unless there was some long-standing animosity between Roberts and Wilmot, I think Wilmot's angry reaction to Robert's article simply reflects the way ownership of a great auk egg or skin bestowed a sense of entitlement. Being overlooked, intentionally or not, was a snub; a failure to acknowledge the prestige that owning a great auk relic conferred. That was, after all, what it was all about.

Dismissive of these exchanges, Newton relegated them to a footnote in his own paper of 1861 recounting his and Wolley's researches in Iceland.[16]

Probably to Newton's consternation, another great auk catalogue appeared in 1868. This time it was by a Swiss zoologist named Victor Fatio. Like many nineteenth-century zoologists, Fatio studied a wide range of topics, but his main work was a six-volume account of the fish, amphibians, reptiles, birds and mammals of Switzerland, published between 1869 and 1904.[17] Some time between 1868 and 1870, and presumably planning to update his catalogue, Fatio wrote to Newton asking him to provide some further information on the subject. Newton said he was happy to help, but added that since his material 'may not be void of interest in this country also', he had decided to publish what he knew, rather than simply sending Fatio the relevant information. Newton may have genuinely been trying to be helpful, but, knowing what we know of his character, it may also have been a clever way for Newton to assert his ownership, if not his priority (as he saw it), of information about the great auk. A lust for status.

Newton's paper – of which Rowley is conspicuously *not* a co-author – was published in the journal *Ibis* in 1870. He emphasised how difficult it had been to compile a perfect, complete list of the skins, bones and eggs that were held in collections. 'For more than ten years, as many of my friends know, it has been my object to do this and, ably assisted of late by Mr G. D. Rowley, I have accumulated a vast mass of materials in my endeavours to trace the history of each specimen.' The difficulty he and Rowley had encountered in completing this task was due, he said, to a small number of ornithologists who were reluctant to provide details of their collections. As we will see later, such behaviour persists to this day.

Newton's comprehensive summary of the whereabouts of 71 or 72 skins, 9 skeletons, the detached bones from 38 or 41 different birds, and 65 great auk eggs placed him – for now, at least – exactly where he wanted to be: back on top of the great auk hierarchy.

Years after Rowley's death, his great-grandson, Peter Rowley, found a letter that Newton had written to Rowley in the 1870s tucked into the back of volume II of the *Alciana*. 'I am sorry to hear of your withdrawal from the garefowl [great auk] undertaking – especially when you ground it on the supposition that you cannot hope to live until I have leisure to devote myself to it.'

Clearly, Rowley was tired of waiting for Newton to start doing something with the mass of facts they had both accumulated, and to produce jointly an 'authoritative book on the great auk'.[18] He had pulled out.

There were probably several reasons why Rowley dissociated himself from Newton. The first was the very real risk that Newton would not include him as a co-author on the monograph. Certainly, Rowley's name did not appear on anything else that Newton published on this topic. Although Newton was a procrastinator and a perfectionist, he was far from idle, since, during the decade between 1860 and 1870, he wrote four lengthy (not to say long-winded) articles on the great auk on a range of topics, each of which might have made a chapter of the much-dreamed-of monograph.[19]

My guess is that their collaboration was doomed from the start; Newton was meticulous and desperately slow, whereas Rowley was 'a scattered disorganised personality'[20] – hardly a good basis for a productive partnership. Simply amassing information on an extinct species doesn't make a book, and it sounds as though Rowley lacked the broad view that Newton undoubtedly possessed, one that would have allowed him to bring it all together. I wonder whether Newton really wanted to share the great auk with Rowley, as he clearly would have done with his friend Wolley, who, after all, was the true originator of the great auk project.

Abandoned, Newton soldiered on alone, continuing to amass material, and continuing to delay putting pen to paper. In 1885, his ambition expired in a despairing gasp of disbelief at the unexpected appearance of a great auk monograph by the almost unknown Symington Grieve. The publication of this impressive volume – *The Great Auk or Garefowl* – was cataclysmic for Newton, not least because it forced him to endure two ironies.

The first was that he was asked by the editor of the journal *Nature* to review Grieve's book. He tried to decline, but in the

Symington Grieve, author of *The Great Auk or Garefowl*. (From Fuller, 1999)

end accepted, only to use the review to vent his utter frustration at not writing the monograph himself.[21] The second irony was one particular sentence that appears in the Preface to Grieve's volume:

> *There is one gentleman to whom the author is under greater obligations than any other, and he is the friend who has made the translations, revised the manuscript, and then proofs, but at his own request will be nameless.*

This was certainly not Alfred Newton. In fact, 'the friend' who helped Grieve was none other than George Dawson Rowley. I cannot help feeling that, having been frustrated by Newton, Rowley thought 'stuff it, I'll share everything I know with Grieve'.[22]

It is little wonder that Newton felt so angry and conflicted when writing his review for *Nature*. Procrastination, fuelled by the need for absolute perfection, had cost him dearly.

Dawson Rowley had died back in November 1878, the same day as his father. Dawson's son, Fydell, inherited thousands of acres, three mansions and all his father's specimens, including the two stuffed great auks and six eggs. When Fydell himself died in 1934, this vast assemblage of specimens was put up for sale at Stevens's Auction Rooms in London. Stevens's specialised in the sale of biological specimens such as birds' eggs and skins as well as butterflies, antiquities and curiosities. This was where the relatives of deceased collectors sent their inherited collections to be auctioned, and it was there, on 14 November 1934, that Vivian Hewitt, the Reverend F. C. R. Jourdain, F. G. Lupton, Harold Gowland and many others gathered, in the hope of acquiring some of George Dawson Rowley's great auk relics.[23]

Chapter 5

All Things from Eggs

To complete his repair of the Bowman Labrey Egg, Graham Axon applied acrylic paint using a fine brush. Because the egg had been so badly damaged, reconstruction required filling in those areas where there was no shell. To do this, Graham implanted fragments of goose eggshell of exactly the right curvature. Then, with considerable skill, he painted these fragments of white eggshell to match the colours and markings on the rest of the egg.

Fortunately for Graham, this was one of the easier great auk eggs to paint. The ground colour is off-white and the markings black and grey. Some of the other great auk eggs are more brightly coloured, with a complex patina that would have required even more skill to replicate.

It was that variation in the colour and markings that helped make great auk eggs both individually recognisable and irresistible to collectors. Up until the 1850s it would have been assumed that it was God who had blessed great auk eggs with their attractive colours, specifically to tempt oologists and to help them identify each egg. In much the same way, people also believed that the all-wise Almighty had varied the coat colour of cattle 'so that every man shall know his own beast'.[1] Darwin put paid to such ideas in 1859 by showing that natural selection provided a far more compelling and testable explanation.

The fact that each great auk egg has a name not only makes them easier to talk and write about, but it also adds a special lustre to their significance. When people started to compile lists of all the extant auk specimens, those eggs were referred to by the lister's name and their number. Symington Grieve in

1885 was among those who did this, so an egg might be referred to as 'Grieve's 69', for example. Things started to become unwieldy as a succession of listers, including Edward Bidwell in the 1890s, Thomas Parkin in 1911 and Paule Marie Louise and John Whitaker Tomkinson in 1966, created their own numbering systems. Anyone wishing to discuss a particular egg had no option but to mention *all* previous names and numbers if they were to avoid confusion. However, while writing his great auk compendium in the 1990s, Errol Fuller decided to give each egg a new name, carefully chosen from some aspect of its history. Not only do Fuller's names provide an easy form of identification, they also bestow a special cultural or even spiritual significance on each egg. The beautiful, red-streaked egg known as John Wolley's Egg commemorates his visionary great auk research in Iceland. The same is true of Alfred Newton's Egg, Vivian Hewitt's Egg, and Dawson Rowley's Egg. There are others, though, like Captain Cook's Egg or Jack Gibson's Egg, whose names are now suggestive of skulduggery and sharp practice.[2]

The colour, the shape and the size of great auk eggs were what oologists coveted. But for a long time it was assumed that these features had little to tell us about the biology of the birds that laid them.

In his monograph, Errol Fuller wrote that:

> ... *all the frantic activity of these men* [who collected great auk eggs] *was, perhaps, to no real point. Alfred Newton ... believed that little of scientific value could be gleaned from the study of a collection. Once assembled, such collections were – he thought – simply things of aesthetic beauty.*
>
> *The few Great Auk eggs that remain must be considered in the same way. What knowledge there is to be drawn from them has – very likely – already*

*been taken. Ornithologists may fuss over what they consider to be significant
'scientific' data, but none of this really matters.*[3]

Fuller's less than optimistic opinion of the scientific value of
birds' eggs was partly a consequence of there having been
relatively little research on eggs since Newton's day, and partly
because the Protection of Birds Act of 1954 had made eggs
politically inappropriate objects of study. Fuller's comment on
Newton is a reference to the preoccupation of Victorian
ornithologists with taxonomy, and Newton's own reluctant
admission that eggs had little to offer the taxonomist.

By the early 2000s sufficient time had passed to allow a
renaissance in research on birds' eggs. Science and technology
had moved on. Armed with current ideas and novel
techniques, scientists had new ways of looking at old eggs.
One discovery was that the key features of birds' eggs – their
shape, colour and size – had not arisen by accident. Rather,
it was now realised that they had evolved in response to the
bird's physical form and lifestyle.[4] This meant that as far as
the great auk was concerned, knowing what has shaped and
coloured the eggs of *other* auks made it possible to infer
something about *its* lifestyle, a lifestyle we cannot now
witness for ourselves.

A particularly striking and beautiful great auk egg is the
Green-blotched St Malo Egg.[5] This egg is extraordinary, not
just because of its extensive green markings, but also because
– having examined thousands of common and Brünnich's
guillemot eggs, as well as razorbill eggs, both in the field and
in museums – I have never seen *green* maculation in any of
them. True, some guillemot eggs have a green ground colour,
but not green markings.

At the risk of allowing the science to eclipse the magic, there may be a simple explanation for the St Malo Egg. The colours of birds' eggs are created by just two pigments: a reddish one known as protoporphyrin IX, and a greenish one known as biliverdin. Mixed in different proportions and applied in different concentrations, these two pigments create the entire panoply of avian egg colours. They are under genetic and physiological control, and sometimes something goes wrong such that only one of the two colours is available. When this happens in birds such as ravens, lapwings and guillemots, it is often the biliverdin (green pigment) that is lacking, giving rise to an egg with reddish markings. Oologists once slavered over these so-called 'erythristic' eggs. More rarely, the protoporphyrin IX (red-brown pigment) is missing, and I think that this may have been what happened in the female that produced the St Malo Egg, resulting in its green maculation. If one was ever brave (or foolish) enough to remove a tiny piece of the eggshell from this particular egg, there are chemical tests that could confirm (or deny) the absence of protoporphyrin IX.[6]

In some cases, *both* pigments may be absent, resulting in a completely white egg. This may be what happened with a great auk egg known as Lady Cust's Egg. Mary Anne Cust (*née* Boode) – probably the only female oologist in the great auk story – was a naturalist, illustrator, cat lover, and friend of Darwin's nemesis, the anti-evolutionary anatomist Richard Owen.[7] Mary's father, Lewis William Boode, made a fortune in the West Indies from his plantations and slaves; he died in the year following his daughter's birth. His widow then bought Mockbeggar Hall in Cheshire, which she re-christened Leasowe Castle, and it was there that Mary grew up. Lady Cust, as she became, possessed a collection of eggs of British birds that was considered 'superior to that held by the British

Museum'. This was a consequence of her having the unreserved assistance of William Yarrell, one of Britain's most eminent ornithologists. Yarrell obtained the white great auk egg for his 'pupil' while in Paris. It eventually ended up as part of Vivian Hewitt's collection – and was one of those that subsequently disappeared.[8]

By comparing the colour of common guillemot and razorbill eggs with those of the great auk, we have a window through which to view an important aspect of the great auk's biology. Eggs of the common guillemot come in a staggering range of ground colours, from white through pale greens and blues to deep turquoise, red and even black, overlain with a huge variety of markings including spots, splotches, streaks, speckles and squiggles. Among the most widely collected eggs of any bird, it was said that no two guillemot eggs are alike. Collectors of stamps, butterflies or birds' eggs lust after unusual and rare types. The men that collected guillemot eggs – and there were many – were prepared to pay high prices for unusual types.

Collectors who coveted guillemot eggs also acquired razorbill eggs, largely because the two species often breed side by side on the same cliff ledges. Slightly smaller and much less pointed than guillemot eggs, razorbill eggs also show a lot of variation in their colour, but the eggs of the two species differ in two striking ways. First, the ground colour of razorbill eggs is usually white or off-white and is never the bright blue or green so common among guillemot eggs. Second, the maculation on razorbill eggs usually consists of spots and smears. While there is some overlap, the smearing typical of many razorbill eggs is extremely rare among guillemot eggs, and the pencilling – long, fine squiggled lines – seen on certain guillemot (and great auk) eggs is almost unknown in razorbill eggs. Despite this difference in the eggs of the two species, I have sometimes found razorbill eggs in museum collections labelled as guillemot eggs, and *vice versa*. They are most likely errors in identification or labelling.

However, when an egg of one of these species *genuinely* has the colour and marking of the other, the explanation may lie in a remarkable record of an apparent hybridisation between a razorbill and a common guillemot.[9] The existence of this bird suggests that hybridisation events in the past may have resulted in these two species sharing a few of each other's genes. Evolutionary biologists refer to the transfer of genetic material between species through hybridisation as 'introgression', and there is good genetic evidence for introgression between Brünnich's guillemots and common guillemots. No one so far has looked for evidence of introgression between common guillemots and razorbills, but the existence of a probable hybrid makes this possible. I wonder whether those genuine, but very rare, cases of common guillemot eggs with razorbill colour and maculation characteristics, and *vice versa*, may be the result of gene exchange between the two species.[10]

Great auk eggs – judging from those still available for us to examine – varied enormously in colour and markings. Such biological variation does not usually arise without reason. In the 1950s and 1960s, a Swiss behavioural scientist named Beat Tschanz decided to investigate why there was so much variation among the eggs of the common guillemot. The brilliant coloration that made many guillemot eggs so conspicuous on the cliff ledge precluded the possibility – suggested by some oologists – that their colour was for camouflage. The more obvious explanation, as Tschanz demonstrated, was that it was for individual identification.[11]

By contrast, razorbills do not seem to be able to identify their own egg. And why would they? Unlike guillemots, they breed well spaced out from others of their own kind, in individual sites, so it is almost impossible for the eggs of different parents to become mixed up.[12]

Now to the great auk's eggs. The variation that they exhibit was quite sufficient for parent birds to identify their own. This is likely to have been important because it seems that great auks bred in densely packed colonies, just as guillemots do. There are several lines of evidence for this. First, the sheer numbers of great auks on tiny Funk Island implies that the birds must have been crowded together when breeding. Second is an account from 1718 by the surveyor William Taverner:

> *The Penoguine Islands* [the Funks] *are in Summer time covered with fowle, of that Name, they are as large as any Tame Goose, their wings are soe small that they can never fly... In the mo* [month] *of June, they come to those islands, which are flat, on which they lay their Eggs, the french from Placentia* [the capital of French-held Newfoundland] *did yearly goe to those islands, & load Boats of 20 Tunns with their Eggs ... they told me that a man could not goe ashoar upon those islands without Boots, for otherwise they would spoile his Leggs, that they were Intirely covred with those fowles, soe close that a Mann could not put his foot between them.*[13]

For birds breeding at high density and incubating their single egg on bare rock with no nest, the ability to identify their own egg would have been important.[14]

And there's one final point relating to the colour of great auk eggs. It appears that, like guillemots, female great auks produced an egg of similar colour and markings each year. This is based on two great auk eggs taken from the island of Eldey in 1840 and 1841 and assumed to have been laid by the same female. Not only are these two eggs incredibly similar both in shape and in their unusual colour and markings, they also both possess a curious and extremely rare spiral deformation of their pointed end, caused, I presume, by a quirk of the female's uterus.[15]

In 1919, the ornithologist Arthur Cleveland Bent, in his book *Life Histories of North American Diving Birds*, described the shape of the great auk's egg as 'almost ovate–pyriform', suggesting 'in general appearance, a large murre's [guillemot's] egg'. For well over a century, oologists referred to the eggs of different bird species using terms like 'elliptical', 'ovate' and 'pyriform', as though these were well-defined and distinct categories. They aren't, but they do provide a convenient, albeit qualitative, idea of the shape of eggs. Later, though, biologists wanted something rather less vague and more quantitative so they invented some indices that would capture the shapes of different birds' eggs.[16]

The beautifully smooth curving symmetry of birds' eggs has intrigued biologists for years. One of the pioneers in this area of research was the Scottish biologist D'Arcy Wentworth Thompson. In his landmark volume *On Growth and Form*, first published in 1917, Thompson showed that despite their subtlety, the shapes of most birds' eggs could be described relatively easily in mathematical terms. An exception, however, was the guillemot. This 'species' extremely-pointed-at-one-end egg-shape defied Thompson's considerable mathematical talents.[17]

Almost a century on, I too wanted to find a way to quantify the shapes of guillemots' eggs. Over the decades that I had studied guillemots, I had seen thousands of their eggs, but never seriously considered making a specific study of their shape. Then, in 2012, a comment on a wildlife television programme explaining – incorrectly – the reason for the guillemot's pyriform egg triggered a challenge.

Re-reading all the accounts I could find about guillemot eggs made me realise that there was much that was unresolved. Especially unconvincing were the various explanations for why guillemot eggs are pyriform. But if I was going to investigate the purpose of a pyriform egg, it was essential that I had a way to describe its shape.

I was fortunate to have a colleague across the road in the Maths Department who could help. When I arrived as a young lecturer in Sheffield in 1976, I was pleased to discover that the University's Department of Statistics was happy to offer advice to anyone undertaking any kind of analysis. It was in this way that I met John Biggins, who, it turned out, had also just arrived from Oxford. We became friends and, over most of my career at Sheffield, he provided me with sound statistical advice and analyses.

John was exceptional in his ability to quickly see to the very heart of any biological question. I would come away from our meetings humbled by his penetrating insights, but full of new ideas as a result. It wasn't always easy; biologists and mathematicians speak very different languages, and we worked at very different speeds. I am what psychologists refer to as a Type A (cursed with a sense of urgency and competitiveness), and John is Type B (painstaking and blessed with patience), which he told me was typical of mathematicians. Over 40 years, ours was a productive partnership.[18]

Having decided to study the eggs of guillemots and other birds, I felt a new world opening up. With Skomer Island as the perfect study location there was a myriad of possibilities. I was also extremely fortunate in having at that time a very capable research assistant, Jamie Thompson. Examining intact guillemot eggs (rather than empty museum specimens) meant climbing down vertical cliff faces to the ledges on which the guillemots bred. Jamie had never done any rock-climbing before, but he was a rapid learner and before long I was placing my life in his hands as he set up the ropes that allowed us to access the guillemots' otherwise inaccessible ledges.

We measured and carefully photographed the eggs on the breeding ledges in the knowledge that it was possible – in principle at least – to derive the eggs' shape from those images. What I did not know at the time was how to do it.

I asked John if he could help, but he said no. He was by this time Head of Department, and busy with much more demanding challenges. There were other mathematicians in the university, but they were neither as approachable nor particularly interested. Luckily for me, the egg-shape question I originally put to John had burrowed quietly into his brain, and before too long we were collaborating. He read all the papers previously published on egg-shape, eventually announcing that 'someone called Preston has already done what you want'. I told him I had read Preston's papers many years ago, and again more recently, but failed to understand them, and was unable to see that he had provided a way of quantifying egg-shape. I was not alone for, as John explained, no one else seemed to have recognised what Preston had done either.

A very clever engineer and amateur ornithologist, Frank Preston created a series of equations that described the shape of *all* birds' eggs. Genius! Sadly though, his equations proved impenetrable to most ornithologists, including those that purportedly had some mathematical know-how. An unfortunate consequence of this impenetrability was that over the following years, a succession of biologists had scrabbled to reinvent a wheel that Preston had already created in the 1950s.

Our task, then, was to make Preston's ingenious methods more accessible – which is exactly what John did. Preston died in 1989, and an obituary described him as bringing 'a penetrating intellect to every subject he touched'. John was the same, and I suspect that he and Preston would have got on – if only I'd started thinking about eggs earlier in my career.[19]

The ultimate goal of my guillemot egg research was to find a convincing explanation for their pyriform shape. It turned out that an egg of this shape was particularly stable on the bare rock ledge (with no nest) on which it was incubated, especially if the ledge was sloping, as it often was. Simply put, the long flat surface of the more pointed half of the guillemot's egg meant that there was more egg in contact with the ledge than there would be with a more curved egg, like that of a razorbill. That extra contact provided the friction to keep the egg relatively stable.[20]

Using Preston's methods, we were able to show quantitatively that the great auk's egg was almost as pyriform as a guillemot's. Like a guillemot, the great auk – as stated very clearly by Martin Martin in 1698 – was thought to have a single, centrally placed brood patch. This led me and many others to assume that the guillemot and the great auk incubated in the same way, standing semi-upright over their similarly pyriform single egg. Indeed, some of the mounted museum specimens – skin and egg together – are positioned in just this way.

To celebrate our success at measuring the shape of its egg, I wrote a paper based on the conventional wisdom of the great auk's single brood patch and semi-upright incubation. No sooner was it published than I began to have doubts. Could I trust the gospel according to Martin Martin that the great auk had only one brood patch when its closest relative, the razorbill, has *two*?

To pursue this, I contacted a curator at a national museum that had several mounted great auk specimens. My request was simple and carried with it the potential to discover something new. I asked if I could examine one of their stuffed great auk specimens to establish whether there was just one or, potentially more interestingly, two brood patches?

As its name suggests, the brood patch is an area of bare skin against which a bird's egg or eggs are incubated or brooded, hidden from view beneath the surrounding feathers. Before the egg is laid and incubation begins, feathers are shed from the brood patch so that its rich supply of blood vessels can transfer heat directly from the parent's skin to the eggs. Some birds have just a single patch, some have two, and some birds, including gulls, have three.

What I imagined I would do with a mounted museum specimen was to gently part the feathers on the lower abdomen to see if there was a single central patch, as Martin Martin so confidently asserted.

The response I received from the curator, however, was disappointing. A very definite 'no', with the implication that my request was inappropriate as the specimen was too valuable for such an invasive examination.

Dispirited, I wrote to tell a German colleague, Karl Schulze-Hagen, what had happened. Karl is a no-nonsense kind of chap – a medic by day and an accomplished, well-connected ornithologist in his spare time. He immediately called up his friend Jürgen Fiebig, a curator at Berlin's Natural History Museum.

Berlin has a single, mounted great auk, one of those killed on Eldey in Iceland in June 1830. Its acquisition came about like this: on 28 December 1830, Professor Johan Christopher Hagemann Reinhardt, director of Copenhagen's newly founded Royal Museum for Natural History – through whose hands most great auk specimens from Iceland passed – wrote to Berlin's curator, Martin Lichtenstein, to tell him that he had a surplus specimen. 'The skin is good, the feathers tight, the bare parts complete ... it is obvious that the skin was stretched when the bird was skinned by a fisherman from Iceland, but this may be corrected when the bird is stuffed.' He added 'I cannot dispose of it for less than 24 Prussian Reichthalers', the equivalent of €1,000 (£850) today. The deal was done; Lichtenstein obtained his iconic specimen in March 1831, and it has been in Berlin ever since.[21]

Without further ado, Jürgen lifted the glass cover from the auk, and started to search for the supposed central brood patch between the bird's legs. Nothing! Moving up the bird's flanks and carefully folding back the feathers, he found a feather-free patch of skin. And there, on the other side, was another. Two brood patches.

On 13 February 2021, Jürgen emailed me to say:

> I am very pleased that I was able to confirm your assumptions with my short analysis … I am always surprised by the simple questions that have not been investigated in detail so far … Of course, one must now examine further Great Auk specimens and perhaps also other related species for clearly visible brood spots.

This news made my day.

Despite the Covid pandemic, Jürgen was subsequently able to examine no fewer than seven other great auk specimens in German museums, all of which, he showed, had two brood patches.[22]

How could Martin Martin have been so wrong? For many amateur and professional researchers alike, Martin is the great auk super-hero; the ultimate trustworthy authority, whose information – gleaned first-hand from St Kildans with great knowledge of the birds they relied on – seemed irrefutable. Few, it seems, ever doubted Martin's account, but while I was writing this, I came across a comment by John Gurney, a perceptive nineteenth-century ornithologist, who queried Martin's assertion about the great auk's single brood patch, reaffirming my belief in the value of checking the older literature.[23]

One of the great auk specimens that Jürgen examined was what is now known as Naumann's Auk, in the Naumann Museum in Köthen, near Leipzig. Like the Berlin bird, this one also arrived via Copenhagen in late 1830 or early 1831 from the natural history dealer Johann Heinrich Frank, whose Danish or German contact wrote:

> *In the month of June 1830, some Icelanders ventured to the Geirfugl skerries at my urging and request … they found there only eight individuals of this rare bird, of which they were able to kill four; of these, two came into my possession, one to Copenhagen, the other is for you* [*i.e.* J. H. Frank].[24]

Six months after the bird had waddled ashore on the Icelandic skerry, its disembodied, flaccid skin was in Naumann's hands.

Johann Friedrich Naumann was central Europe's best-known and most respected nineteenth-century ornithologist. Intrigued to see what Naumann had written about the preparation of this particular great auk skin, Karl Schulze-Hagen started to investigate. The number of articles, papers and books on the great auk is so vast that it was almost impossible to imagine that anything new could be discovered, but Karl found something extraordinary.

Naumann, the son of the farmer and bird enthusiast Johann Andreas Naumann, enjoyed just four years of formal schooling. Despite working long hours on the family farm in the tiny hamlet of Ziebigk in the heart of rural Saxony-Anhalt, he was to become Germany's most celebrated ornithologist.

Between 1795 and 1817, Naumann's father had published a four-volume handbook of birds, the title of which translates as *Natural History of the Land and Water Birds of Northern Germany and Neighbouring Countries*.[25] Johann had helped his father with both the text and the illustrations, and the book proved to be so popular that Johann Friedrich subsequently decided to produce an even more comprehensive account of Germany's birds. Acknowledging the central role his father had played, Johann Friedrich entitled this book *Johann Andreas Naumann's Birds of Germany*, even though it was largely his own work. Published between 1820 and 1844, this 12-volume work quickly became central Europe's definitive ornithology text, and it remained so for more than a century.

In writing the book, Johann Friedrich struggled with species he had not seen or handled himself. Living so far from the coast meant that seabirds were a particular challenge. Seeking assistance, he wrote to friends who had visited seabird colonies and whose collections included seabird specimens. Several ornithologists living in Schleswig-Holstein in northern Germany – an area bordering both the Baltic and the North Sea – were especially helpful, inviting him to visit the gull colonies there in 1819. In 1840, as he began to prepare the 12th and final volume of his *magnum opus*, Naumann travelled to the island of Heligoland, then British-held but off the coast of Germany and Denmark, specifically to see and sketch the guillemots, razorbills and puffins that bred there. Perhaps surprisingly, Naumann also intended to include an account of the great auk. He did so because it was considered a German bird, having been recorded (albeit just once) in Kiel harbour, where one was shot in 1794 or 1796.

To write his account, however, Naumann needed a specimen. After great auks were exterminated in the north-west Atlantic, their only known breeding colonies were – as we have seen – in Iceland, which at that time was under

Danish control. Conflict between Denmark and Britain in the early 1800s meant that almost all great auk specimens leaving Iceland passed through Copenhagen, and they were both rare and expensive. Naumann was offered one in 1824 in exchange for the skin of a lammergeier (a species now also known as a bearded vulture), but that too was extremely rare and he could not afford the swap.

Finally, in the winter of 1830–31, he obtained a specimen through Johann Heinrich Frank. Frank had obtained the bird from another dealer, probably Carl Franz Siemsen. We do not know if Naumann paid Frank cash or whether, as seems more likely, he exchanged it for the skin or skins of some other rare birds. Remarkably, Naumann's Auk was killed during the very same raid on Eldey as the Berlin Auk mentioned earlier (see p. 104), and the two birds may even have been a pair.[26]

With a specimen of his own, Naumann was able to produce a description. His thoroughness and expertise were such that he soon noticed features in the great auk that no other ornithologist had commented on. However, like my colleague Jürgen Fiebig, Naumann was wary of drawing conclusions from a single specimen, and with Frank's help he was eventually able to examine eight or nine skins. Among these was an adult great auk in winter plumage – one of only two winter-plumage skins known, and now recognised as the bird killed by an Inuit kayaker during the winter of 1814–15 or 1815–16 (see p. 50).[27]

Naumann's description covered the plumage, measurements of the bird's body length, and its wings, beak, feet and toes. He also noticed that the specimens said to be female were slightly smaller than the supposed males, but no one had kept adequate records of the sex of specimens based on their internal gonads. Examination of the feet allowed Naumann to draw some inferences about the bird's body posture and gait on land, drawing careful comparisons with the razorbill.

Tab 337.

ALCA impennis *Flugloser Alk.*
1. M. Hochzeitkl. 2 Jugendkl.

Johann Friedrich Naumann's illustration of great auks in winter (back) and summer (front) plumages. This was completed in 1844, the year the last two great auks died. (Courtesy Klaus Nigge and Naumann Museum, Köthen)

In his account of 1844, he writes:

> *The feet are not very large, but strong, almost plump … [and] lie even farther back* [than in the razorbill] *… The heel joint is particularly strong, and the sole of the foot so broadly pressed from the joint to the toes that this unmistakably indicates the bird stands and walks on this part of the foot more than its relatives. These soles, both of the legs and the toes, are rough … and have little warts.*

Making his comparisons between the great auk and the razorbill, Naumann was surprised by the number of differences he discovered, suggesting that rather than the two species being members of the same genus, *Alca*, the great auk ought to be in a genus of its own.[28]

He also stated that both 'male and female [great auks] incubate alternately, which is proved by the brood patches which both possess, one on each side of the abdomen, as in *Alca torda* [the razorbill]'. His comment that the great auk had *two* brood patches is remarkable, both for its matter-of-factness and the fact that prior to Karl Schulze-Hagen's rediscovery of Naumann's account, no one (certainly in recent times) had thought it worth reading.

Two brood patches, then – one on either side – but *how* did a great auk incubate its egg? With no written descriptions, the only way to find out was to look at how the razorbill does it. I thought I knew the answer to this, for I had watched razorbills incubating for decades. But the more I thought about it, the less confident I felt. It was the time of the Covid lockdown and Skomer was inaccessible, so I headed off to Bempton Cliffs on the Yorkshire coast to watch and film incubating razorbills – *really* looking this time. It seems so simple now, as I write this, but it took a lot of looking to see that the egg is held (completely out of sight) beneath one wing, on the outside of the leg against the brood patch. Having two brood patches means that razorbills can incubate facing either way, while keeping the egg adjacent to the safety of a rock wall. Two brood patches also meant that the great auk, like the razorbill, almost certainly incubated in a horizontal position, rather than in the guillemot's semi-upright posture.

A great auk positioning its egg with the ventral edge of its lower mandible (left) prior to settling down to incubate (right). (Illustrations by David Quinn)

One further point. My artist friend and great-auk illustrator David Quinn told me that the curve on the end of the great auk's lower mandible exactly mirrors the shape of the blunt end of the egg. The great auk would have moved its egg into position under its wing using its bill, and the shape of the lower mandible may thus be an adaptation for this purpose.

My assessment that the great auk incubated in a horizontal position is consistent with statements made by those intent on catching great auks, that the birds were 'taken by surprising them where they sleep'. Presumably the men thought that since the birds were lying down and horizontal they were asleep.[29]

I subsequently discovered that there is possibly a first-hand observation that confirms the great auk's incubation posture. An Icelandic farmer, probably one of those involved in the great auk raids, described the great auk as *laying* on its egg – that is, like a razorbill, horizontally. Few (or no) others that visited great auk colonies either in Newfoundland or Iceland bothered to note the behaviour of undisturbed birds. As the raiders approached, most birds would have been standing up from their eggs, ready to run.[30]

Everything we know about the great auk indicates that it bred very close to its neighbours, much like the guillemot. To breed like this, pairs cannot be too choosy about the substrate on which they lay their egg. Proximity, and thus safety from predators, was the priority. This in turn means being relaxed about the substrate, and a pyriform egg gives guillemots – and presumably gave great auks too – the flexibility to incubate their precious egg even on sloping terrain.

After several weeks of incubation, the great auk chick would have chipped its way free from its calcareous shell. This new phase in the great auk breeding cycle is one we know least about. It was neither seen nor recorded by anyone. But it is one that, with the benefit of hindsight and current knowledge of other auks, we can have a good go at reconstructing.

Chapter 6

The Chick That Never Was

Missing eggs, missing chicks; so much of the great auk's life is missing. Not just the 13 eggs that Vivian Hewitt once owned, but the bird's entire repertoire of behaviours, and most significantly, any description of a great auk chick.

Given the exalted position in which great auk eggs and skins were (and still are) held, it is curious that there are no skins of great auk chicks – anywhere. Were one to come onto the market today, the scramble for acquisition would be brisk and the price exorbitant. To understand the absence of great auk chick specimens, and the implications of that vacuum, we need to return to Funk Island during its peaceful, pre-human invasion era.

One day, from within the still intact eggshell, the chick begins to call. It is like the incessant bleeping of a mobile phone that the parents cannot ignore. Standing up and over the egg, they respond with low rumbling growls. The following morning, a small hole appears in the large end of the egg, through which the white egg-tooth on the tip of the chick's beak is visible. The calls are more insistent. Two full days later, after more chipping away by the chick, the eggshell finally breaks apart and the young great auk is free. Despite its wide-open eyes, the newly hatched chick is covered in damp down and is not particularly prepossessing. Barely able to stand, the chick crouches on its rocky bed until one of its parents pushes the chick under its wing where it can dry off and keep warm.

Over the following days, the parents take turns to brood the chick, while the other forages out at sea, returning periodically with a beak clasping small capelin or herring for their hatchling. Adults and chick talk to each other and preen each other incessantly, creating a bond based on recognition and affection. The parents' calls are long drawn-out *rrrrrrr*s, like a baritone tomcat; the chick responds with fruity falsetto peeps. By the time the chick is ready to leave the colony, the atmosphere has reached fever pitch. It feels like a motorcycle engine revving up, ready to race off.

The big day. The seas, for once, are calm with the Atlantic swells washing gently up the sloping shoreline. It is a day of rare sunshine and 20-degree heat, forcing some parent birds to pant with open beaks to stay cool. Despite the unusual warmth, the colony continues in its state of barely suppressed excitement. The colony is like an audience anticipating a rock band's appearance at an open-air concert. Tension builds as the light fades towards darkness. Anticipating their graduation, the great auk chicks squeal with excitement, uttering individually distinct *weeloo weeloo* calls. The parents respond 'I'm here' with reassuring rumbles. The expectation is almost unbearable. Exuberant chicks are jumping on the spot, flapping their tiny wings. The adults calmly stand and wait, calling occasionally and touching the chick with their bill. And then, as though the first rock stars have emerged onto the stage, there's a roar of excitement as the performers brace themselves to begin. The male shuffles away from the spot that has been the nursery, and the chick follows. Growling encouragingly, the male walks towards the sea. With surprising agility, the chick runs ahead and plunges into the waves. It is a vast exodus; almost the entire colony – thousands of fathers and chicks – are on the move. From the air it would look like a football crowd streaming out of the gates at the end of a match. But far from being the end, this is just the beginning. Sticking close to its father's side, the chick joins the vast entourage swimming out into the night. As the chorus of sound on the island dies away, it is replaced by the occasional cries of chicks on the sea anxious to maintain contact with their fathers in the darkness.

By dawn the next day, the seas around Funk are empty. The birds are safely offshore, away from predatory gulls and closer to their feeding grounds on Newfoundland's Grand Banks.

This is my fantasy: a kind of dialogue with the dead, for as far as we know, no one ever witnessed what I have described. It is an informed fantasy, though, based on my knowledge of other auks. I have been careful here not to say anything about how old the great auk chicks were when they left the colony, for this is one of the great great auk mysteries.

Why does it matter? The answer lies in the fact that the auk family – some 22 species in all – shows remarkable variation in the way their chicks develop at the colony. At one extreme, the young of the various puffin species remain in their breeding burrow or crevice until they are about six weeks old, and fledge independently of their parents. Technically, this is referred to as a 'semi-precocial development strategy'.[1]

Lying at the other extreme are the Pacific murrelets, of which the ancient murrelet is the best studied. Their young leave the colony just two days after hatching and are said to be 'precocial'. They are accompanied by both parents.

In between the extremes are three species: the common guillemot, Brünnich's guillemot, and the razorbill, whose chicks all follow an 'intermediate' strategy, leaving the colony at about one quarter of adult weight, aged 17–21 days, and accompanied by their father.[2]

We have the wealthy and well-educated Martin Martin back in 1698 to thank for seeding the idea that great auk chicks might have been precocial, leaving the colony just a few days after hatching. Inspired by the Royal Society's increasing interest in remote human societies, Martin travelled to the island of St Kilda in 1697, where he and his entourage spent an uncomfortable few weeks. Here, on Britain's most remote outpost, 40 miles west of the Outer Hebrides, Martin documented what the residents told him about their extraordinary seabird-dependent lives.[3] During their stay, Martin and his crew were provisioned by the humble residents, but struggled to cope with their daily ration of a barley cake and 18 guillemot eggs. The St Kildans were well used to such a diet, but Martin and his colleagues quickly became monumentally constipated – a condition whose discomfort could be offset only by a soothing honey-based concoction.

Between enemas, Martin took notes from the residents on the different seabirds upon which the resident islanders were dependent. The great auk, he was told, arrived around the first of May and 'goes away about the middle of June' – a period of just six or seven weeks – a much shorter breeding period than any of the other auks, implying that the chick must have left the breeding site soon after hatching. The idea of the precocial great auk chick was born.

Subsequent writers, myself included, accepted this since it so neatly explained another curious fact about the great auk: the absence of *any* sightings of great auk chicks on the islands it bred upon.[4] One such account comes from John Gould, the bird-man darling of wealthy Victorians, who coveted his big, bold, beautifully illustrated bird books. Gould excelled not only as an ornithologist, but also as a hard-nosed businessman. In the mid-1800s, he was considered Britain's foremost bird artist. This is ironic given that it was actually his wife, Elizabeth, and several others, who executed the paintings based on his sketches. Early in his career, Gould was employed as the taxidermist at the Zoological Society of London. This was where Darwin deposited the bird specimens he had collected on the Galápagos islands. Gould recognised that those specimens comprised a unique but closely related group of finches, helping to guide Darwin towards the idea of natural selection.

While Gould was talking finches with Darwin, he was also busy with the lucrative business of producing lavishly illustrated bird books, capitalising on the vogue for such volumes. These tomes were stupendously extravagant. The constant need for subscribers and wealthy patrons – together with an overwhelming desire to maintain his position at the top of the ornithological ladder – seems to have sometimes tempted him to embroider the facts to make his books more appealing.[5]

Among Gould's earliest publishing ventures was the imperial folio, five-volume *Birds of Europe*, published in several parts between 1832 and 1837. Gould wrote the text and the stunning illustrations were executed mainly by his talented wife. Volume 5 covers the auks, including the great auk, about which Gould wrote:

The young ... take to the water immediately after exclusion from the egg, and follow the adults with fearless confidence.

What could be clearer? The great auk chick was obviously precocial. But where did Gould get this information? Writing in 2004, Jeremy Gaskell said that Gould's statement 'appears to bear the hallmarks of an eyewitness account', adding that 'it is doubtful if such an assertion would ever have been made in the absence of at least circumstantial evidence ...'.[6]

But there's no evidence, and no evidence that Gould ever travelled to anywhere where he might have seen a great auk at its breeding colony. Indeed, other parts of Gould's text suggest that he actually knew very little about the great auk:

The seas of polar regions, agitated with storms and covered with immense icebergs, form the congenial habitat of the Great Auk: here it may be said to pass the whole of its existence ... It is found in abundance along the rugged coasts of Labrador; and from the circumstances of its having been seen at Spitzbergen, we may reasonably conclude that its range extended throughout the whole of the Arctic Circle, where it may often be seen tranquilly reposing on masses of floating ice ...[7]

This is nonsense. The great auk was not an Arctic bird, and would rarely have reposed amid masses of ice. Gould created an illusion of authority, embellishing and romanticising some earlier (erroneous) accounts, to make his text more attractive. He does not tell us where he obtained his information, which is fair enough since his books were not mean as 'scientific' treatises, but Martin Martin's account is his most likely source.[8]

With more 'fearless confidence' in Gould than I have, Jeremy Gaskell emphasised Gould's assertion that the chick left its breeding site with both parents – 'the young ... follow the adults'. This, he says, is what one might expect if the recently hatched chick required the care of both parents at sea. It may be true that a recently fledged chick would benefit from the care of both parents, but equally likely is the fact that Gould – obsessed by the Victorian notion of wholesome monogamy – thought that adding an 's' to 'adult' would appeal to his readers.[9]

There is another observation Gaskell considered consistent with the idea of a precocial great auk chick. It concerns the Swedish Count Frederik Raben, and the Danish merchant Frederik Faber, the only ornithologists to get to a great auk breeding colony before the birds became extinct (see p. 66). On 1 July 1821, the two men (with one of the count's servants) visited Geirfuglasker off south-west Iceland. On discovering no sign of great auks at this well-known breeding location, they assumed that the birds must have moved elsewhere as a result of persecution in previous years. Gaskell's interpretation, however, is that there were no great auks present because the chicks had already departed with their parents. It is a possibility. But if that had been the case, then Raben, who got ashore – not without difficulty because of the swell – and climbed up to where great auks were known to have bred previously, would have seen evidence of the recent breeding of this bird. These might have included the leathery inner egg membrane, together with fragments of eggshell that in other auks persist at the breeding site for some time after hatching. In his eager quest for the great auk, the scientifically astute Raben must surely have looked for these clues.[10]

There is actually a very simple explanation for why no one recorded seeing great auk chicks on land. It is because everyone visiting a great auk colony, whether it was Funk Island, Geirfuglasker, Eldey or St Kilda, was there to kill adult birds and collect their eggs. As I mentioned earlier, it was standard practice to trample down the eggs (and chicks) that were present so that any eggs found on the following few days were sure to be fresh. Such is human nature; men did this regardless of whether or not they intended to come back, just to be on the safe side. Egging was a dirty business, as John James Audubon makes clear with his graphic description of visits to common guillemot colonies on the North Shore of the Gulf of St Lawrence in the 1830s:

At the approach of the vile thieves, clouds of birds rise from the rock and fill the air around, wheeling and screaming over their enemies. Yet thousands remain in an erect posture, each covering its single egg … their assassins … walk forward exultingly … See how they crush the chick within its shell, how they trample on every egg in their way with their huge and clumsy boots …

*when they leave of the isle, not an egg that they can find is left
entire... For a week each night is passed in drunkenness and
brawls, until, having reached the last breeding place on the coast,
they return [to] ... collect the fresh eggs ...*[11]

When HMS *Salamine, en route* from the Faroes to Reykjavik in
July 1808, took advantage of calm seas to visit Geirfuglasker,
the crew went ashore 'where they remained the whole day,
killing many birds and *treading down their eggs and young*'.[12]

Clearly, when people visited great auk colonies, few if any
eggs were allowed to hatch. As well as those eggs destroyed
deliberately, the mayhem created by seafarers intent on
capturing adults or picking up eggs would have resulted in a
huge collateral loss of eggs. I can all too easily imagine the
havoc. On a visit to Funk Island, I watched in disbelief through
my binoculars as a journalist who had accompanied us walked
as casually as though through a wildflower meadow into a
group of breeding common guillemots, causing the birds to
flee in panic, scattering their eggs as they fled. On another
occasion, I witnessed the immediate aftermath of a visit by an
Arctic fox to another low-lying island, this time in Labrador,
where guillemots also bred on flat ground. Hundreds of eggs
had rolled away and smashed. Just one or two birds had
returned and were nervously incubating their egg amid the
mass of broken eggshells, a yolky mess covering the rocky
ground.[13]

And there the story hung, unresolved, until I happened to
re-read Martin Martin's account of his stay on St Kilda. Martin
relates how the residents told him that unlike the guillemots
and razorbills, the great auk never laid a replacement if its egg
was lost or taken. As Audubon's eggers and the farmhands that
collected eggs at Bempton Cliffs in Yorkshire knew, guillemots
and razorbills produce a second egg two weeks after their first
is taken away, and sometimes a third if the second is taken. As I

read about the great auk's inability to produce a replacement egg (and assuming that on this particular point Martin was correct), it struck me that this could be the explanation for their exceptionally short stay – arriving in early May and leaving in mid-June – at the colony. If their eggs were taken, as they were by the St Kildans (and by others at colonies elsewhere), and the birds were not programmed to produce a replacement egg, then there were no chicks to be fed, and little point in the adult birds hanging around. They simply left. Suddenly, like a wave sweeping up the shore and washing away a child's sandcastle, the very foundation on which the precocial case had been built was demolished. Time to think again.

In the absence of the chick itself, one way of assessing the great auk's likely chick-rearing strategy is by drawing comparisons with its relatives. Let's start with those precocial Pacific murrelets, whose most striking feature is their huge eggs relative to the female's body weight. An average human baby weighing 3.2kg (7lb) represents about 6 per cent of a mother's weight. The eggs of different bird species vary from 2 to 30 per cent of the female's weight, with the murrelets among the largest at 22 per cent. Unlike humans and other mammals where the upper size-limit of a baby is set by the size of head that can pass through the pelvic girdle, birds have an 'open' pelvis, allowing them to produce relatively enormous eggs. And a relatively large egg is necessary to produce a chick that soon after hatching is sufficiently well developed – that is, precocial – to be able to scramble down to the sea, jump in, swim and dive, accompanied by its parents. It is not just the murrelets; all bird species whose chicks are precocial at hatching, such as ducks, pheasants and kiwis, produce relatively large eggs.

It should be straightforward, then, to decide if great auk chicks were precocial simply by looking at the size of their eggs relative to the mother's body weight. There are enough great auk eggs in museums and reasonable estimates of adult weight to do the calculation. The result is very clear. Great auk eggs are close to the size we would expect if their chicks were *not* precocial.[14]

End of story, job done, you might think, but no. My colleague Tony Gaston, who studied ancient murrelets and their precocial chicks for several years on Haida Gwaii (previously known as the Queen Charlotte Islands), thought differently. He said to me that in terms of the absolute size of its egg (rather than its relative size), there was no reason why a great auk egg could not produce a precocial chick. Adult ancient murrelets weigh around 200 grams (7oz), and their eggs weigh about 45 grams (1.6oz) when fresh. A recently laid great auk egg weighing 351 grams (12.4oz) was, in theory at least, quite large enough to produce a *small* precocial chick. Other auk biologists agreed, and it fitted neatly with Martin Martin's early departure statement. I agreed, too, because at the time Gaston's argument seemed logical.[15]

However, another colleague, Alasdair Houston, a theoretician at the University of Bristol, disagreed. As the saying goes, there is more than one way to skin a cat. Houston's approach to answering the question was to construct a time- and energy-budget model to compare the efficiency of the great auk taking its chick to feeding grounds at sea soon after hatching – *i.e.* being precocial – versus delivering food to the chick at the colony for two or three weeks – *i.e.* being intermediate, like guillemots and the razorbill. Interestingly, this modelling approach was suggested first in the 1830s by Frederik Faber, the friend of the Swedish count who had searched unsuccessfully for great auks in Iceland. Faber decided that the great auk's chick strategy was the same as that of guillemots and the razorbill, that is, intermediate – exactly as Alasdair Houston and his colleagues concluded from their calculations.[16]

Where does that leave Tony Gaston's idea? His suggestion that a great auk's egg was large enough to produce a precocial chick rests on two important but unstated assumptions. The first of these is that the great auk faced neither 'phylogenetic' nor 'allometric' constraints. Let me explain. The phylogenetic constraint is that the great auk's closest relatives, the razorbill and the common and Brünnich's guillemots, all produce 'intermediate' chicks, so it would be odd if the great auk did something very different.[17] The allometric constraint is that, as exemplified by a

A great auk chick accompanied by its father at fledging, as I imagine it. (Illustration by David Quinn)

human baby's body proportions, you cannot simply scale up or down in size, willy-nilly. A baby's head, heart, brain, etc., are not the same *relative* size as those of an adult. If the great auk were to have produced a precocial chick, its egg would have to have been substantially larger and heavier than the estimated 351g.

All in all, the very short breeding season – on which the presumption that they had a precocial chick is based – was a consequence of great auks not re-laying if their egg was taken, and not hanging around thereafter. It therefore seems most likely to me that the great auk chick followed an intermediate developmental strategy, just like its closest relatives, the common guillemot, Brünnich's guillemot and the razorbill, and fledged at two to three weeks of age.

There is one bit of information that could solve for ever the mystery of the great auk chick's development. It relates to the bones of *young* great auks recovered from the soil and guano on Funk Island. Some of these bones were found by Frederic Lucas during his visit there in 1887, and he states that they were subsequently stored in the Smithsonian Institution in Cambridge, Massachusetts. Lucas relates in his report that 'The number of bones from young birds is extremely small.' He adds that 'this all but total lack of them is readily accounted for by the fact that after the merciless slaughter of the Auks had fairly commenced, few, if any, eggs were allowed to hatch'. Although Lucas provides a detailed, quantitative description of the adult bones, including 300 of the many humeri recovered during his expedition, he writes nothing more about the bones of young birds.[18] A century later Keith Hobson and Bill Montevecchi also refer to a single leg bone (a tibiotarsus) of a 'nestling' great auk found on Funk Island.[19]

I was tremendously excited to learn of the existence of these bones, because of their potential to resolve the mystery of the great auk's chick. I imagined three scenarios after being allowed to see these bones. In the first, the bones are tiny and from chicks just two days old. That wouldn't help because those chicks – had they survived – might have fledged at a later age.

In the second, the bones are from chicks aged 35 to 40 days old. If that was the case, I would have to go back to the drawing board because, so far at least, no one has seriously suggested that great auk chicks remained this long at the colony.

In my third scenario, the bones are from chicks aged 10 to 20 days old. This would be exciting since it would be consistent with great auks having an intermediate fledging strategy, like guillemots and the razorbill.

The reality, however, was desperately disappointing. The Smithsonian's curator, Chris Milensky, told me that despite a careful search through the thousands of great auk bones in their collection, he and an osteology expert were unable to find *any* belonging to young birds. I was surprised. Lucas understood the value of these particular specimens, yet in the intervening years they seem to have been lost.

As for the nestling tibiotarsus, 30 years after the event – Keith Hobson and Bill Montevecchi were unable to find even the photograph they had taken of the bone. Stalemate.

Paradoxically, these missing bones highlight, like little else, the value of great auk material in allowing us to reconstruct its biology. In the hierarchy of great auk remains, random bones are at the bottom of the pile. Above them lie complete skeletons, intact eggshells and stuffed specimens, and it is these relics that constitute the bird's afterlife.[20]

PART 2

AFTERLIFE

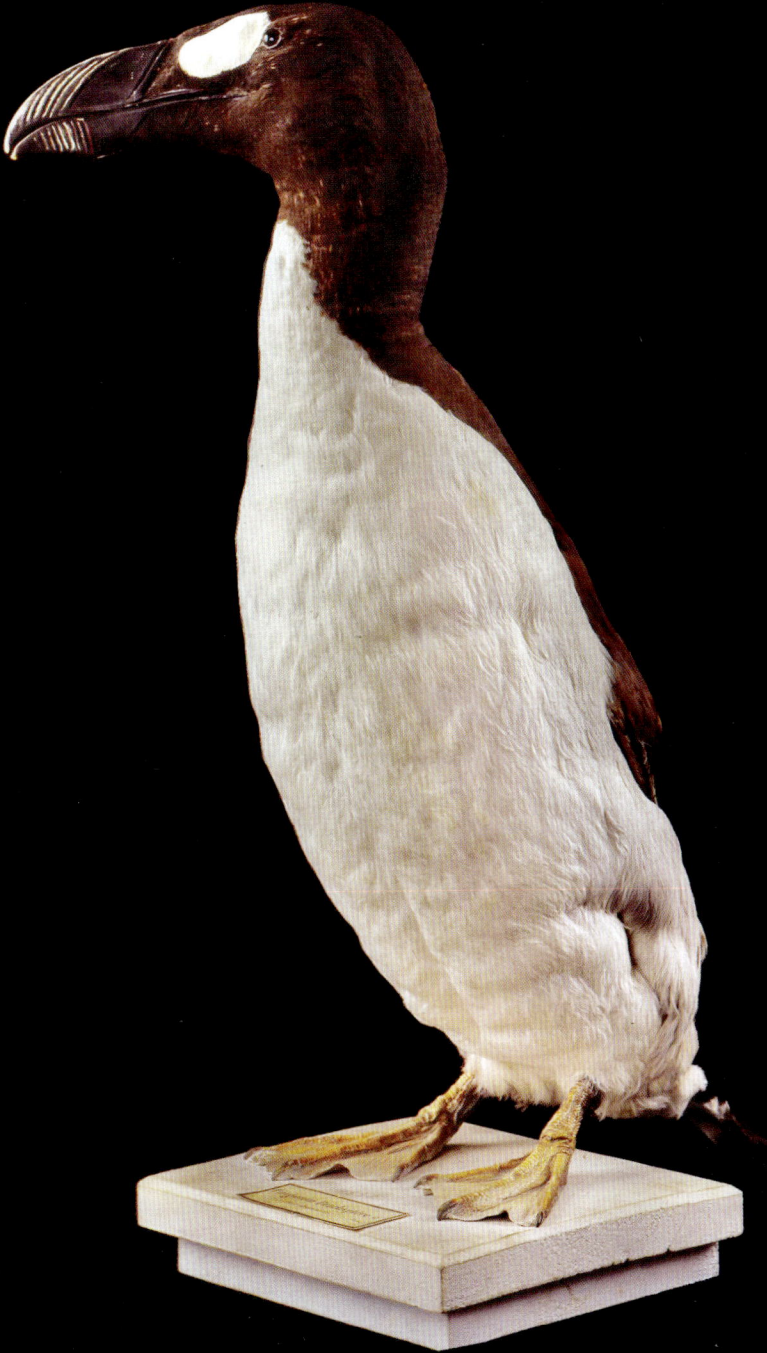

Hoppa's Auk, one of the most lifelike stuffed great auk specimens. Taken on Eldey, probably in 1831; now in a private collection. (Errol Fuller).

Above: Funk Island, an 800m x 300m granite slab off the Newfoundland coast. The area once occupied in summer by great auks was probably the dark, grassy green and white areas combined. These areas are the breeding grounds of common guillemots, puffins and gannets. (Sabina Wilhelm). **Inset left:** The tip of a great auk beak found in the Funk Island soil. (T. R. Birkhead). **Inset right:** Some of the great auk gizzard stones (*c.* 0.5cm diameter) found by Owen Bryant in 1908 in the Funk Island soil. (T. R. Birkhead).

Below: Funk Island, here covered by mainly common guillemots; the cairns are built from rocks used to create the corrals in which great auks were herded and slaughtered. (W. Montevecchi).

Above: The great auk's closest living relatives: **(a):** Brünnich's guillemot (B. Lyon); **(b):** common guillemot (K. Nigge); **(c):** razorbill (K. Nigge).

Below: The Versailles Great Auk(s) painted by Nicholas Robert in 1666–1670. (Österreichische Nationalbibliothek).

Alka Hoieri.
Mergus Americanus Clusij Exot.

Plongeon d'Amerique.

Above: Edward Lear's great auk. A hand-coloured lithograph completed in the early 1830s. (From John Gould's *The Birds of Europe* (1832–1837), courtesy Zoological Society of London).

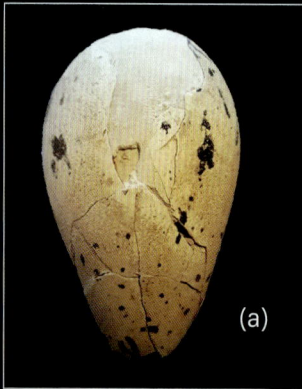

The Bowman Labrey egg, (a) as received by Graham Axon before repair; (b) the dissociated fragments of the egg prior to its restoration; (c) the restored Bowman Labrey great auk egg. (Graham Axon).

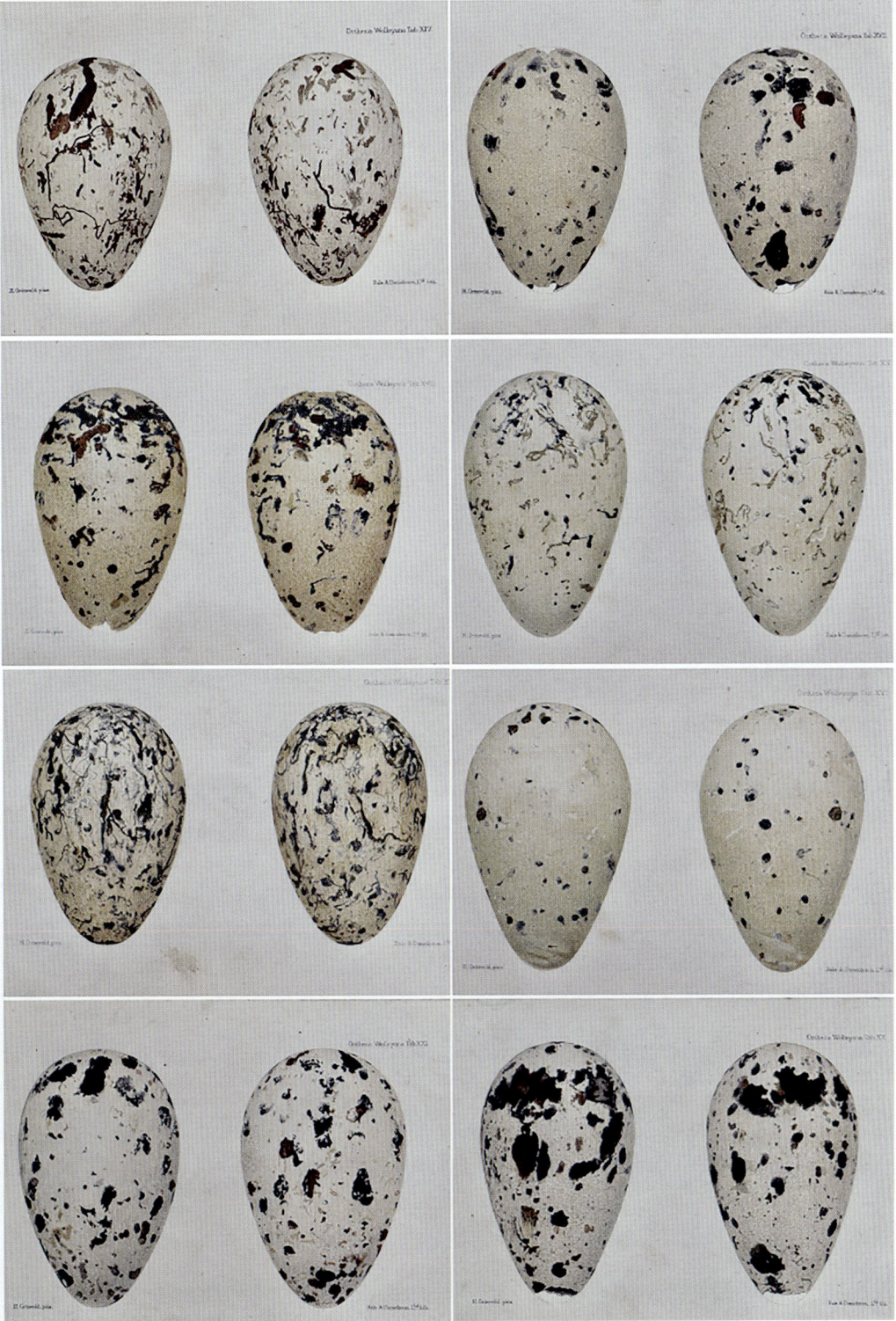

Above: Eggs of the great auk. Paintings of each side of eight eggs by Henrik Grønvold for Alfred Newton's *Ootheca Wolleyana* (1907).

Above: The Berlin Great Auk, with coloured pins showing the location of the bird's two lateral brood patches. (Carola Radtke).

Below: 'Fortress' Bryn Aber and (inset) Hewitt's lagoon. Taken by an unknown photographer in the 1970s, after Vivian Hewitt's death, but while Jack Parry was still living there. (Hewitt Papers).

(a)

(b)

(c)

(d)

Above: The heads of four mounted great auks showing variation in their age, as deduced from the number of grooves on the bill, and the differences in adult summer and winter plumages; **(a):** the Irish Auk, in adult winter plumage (M. Linnie); **(b):** the Copenhagen Winter Auk, an adult in winter plumage (G. Romello, University of Copenhagen); **(c):** the Tunstall Auk, an immature bird (T. R. Birkhead, courtesy D. Gordon); **(d):** Hoppa's Auk, an adult bird in summer plumage (see also plate 1) (Errol Fuller).

Right: An immature (first-year) great auk, as judged by its small beak without furrows and the absence of a white spot in front of the eye, by an unknown 17th century artist.(Courtesy of the Jagiellonian Library, Krakow).

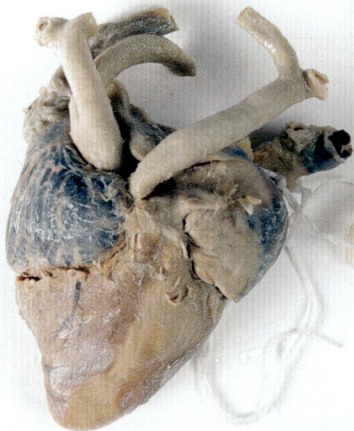

Left: The spirit-preserved heart of one of two last great auks (the male) to be killed. (Natural History Museum of Denmark; courtesy G. Romello, University of Copenhagen).

Alea impennis. ♂
Island. 1844.

Below: Great auks allopreening. A 2023 painting by David Quinn.

Chapter 7

Playboy, Pilot and Ornithologist

The morning of 30 May 1922 was bright and clear. Releasing his motor launch from its moorings in Rhossili on the western tip of the Gower Peninsula in south Wales, 34-year-old Vivian Hewitt and his mechanic friend George Griffiths set a course for the tiny, uninhabited island of Grassholm. Far out in the Irish Sea, Grassholm, like the more inshore islands of Ramsey, Bardsey, Skokholm and Skomer, was famous for its seabirds. Hewitt later wrote in his diary that it was 'one of those hot lazy days when the sea itself appears to be almost asleep'. Even so, arriving at Grassholm seven hours later, the swell was huge, and it was only with luck that Hewitt and Griffiths were able to scramble safely ashore.

The gannets that they had come to see were nesting at the north-west end of the island. 'The stench and uproar of the colony are indescribable. Masses of putrefying fish lay about in all directions, and the smell was almost overpowering.' Hewitt estimated there to be between 800 and 1,000 pairs breeding on the island. 'I found eggs in all stages of incubation, from fresh to chipping out, and young chicks ranging from just hatched … to a fair size.' He noted the gannet's eggs varied 'in shape from thin to broad elongate ovate, and from thin pointed ovals to broad pointed ovals, and in size they range from 59mm to 70mm in length and from 30mm to 35mm in diameter'. Hewitt took some gannet eggs to add to his collection.

After an exciting day, Hewitt and Griffiths left Grassholm at 7:30pm, somewhat later than intended. Motoring against the tide, progress was slow and they soon found themselves travelling in the dark. They struggled to navigate because of a faulty compass, and it wasn't until dawn the next morning that they found themselves back at the Gower. They were fortunate to have travelled when they did, for the following night a fierce storm ripped Hewitt's boat from its moorings and buried it beneath the waves. With no mention of these misadventures, Hewitt published a short account of his trip, describing it as one of the most interesting he had ever made 'in the study of oology'.[1]

Hewitt was not unusual in taking a few eggs as souvenirs. It was what everyone visiting seabird colonies did at the time. However, Hewitt's interest in eggs was more than casual, as we can tell from the way he describes the shape of the gannets' eggs, and his care in reporting the variation in their size. His mention of the 'study of oology' – the egg-collectors' quasi-scientific term for the taking of birds' eggs – tells us he was serious about the subject.[2]

Hewitt's handful of gannet eggs from Grassholm was the beginning of his oological obsession. By the late 1940s his collection, numbering hundreds of thousands of birds' eggs, was greater than that of most museums, including Britain's Natural History Museum. Holding pride of place in Hewitt's collection were *thirteen* eggs of the great auk.

One of these, now lying in state in the National Museum of Wales in Cardiff, bears Hewitt's name. Thought to have come originally from Iceland in the early 1800s, it was taken from there to France by the captain of a fishing vessel, where it was given or sold to a merchant in the French town of Bergues. This unknown merchant in turn is said to have given the egg to a young collector (also unnamed), whose entire collection was bought by a certain Monsieur de Méezemaker around 1817. In 1900 the egg was sold again, this time to an Alfred Vaucher, of Lausanne, from whose son Hewitt bought it in 1937.[3]

Vivian Vaughan Davies Hewitt was born on 11 March 1888 into a wealthy brewing family in a house called Holmefield, in Grimsby on England's east coast. Educated at Harrow School, Hewitt left at the age of 16 to embark on a grand tour of Australia and South Africa. More interested in mechanics than academic study, he was sent on his return to the Marine Engineering Department at Portsmouth Dockyard. From there he became an apprentice locomotive builder at Crewe as a 'privileged apprentice' – the son of a rich man, 'bitten by the railway bug'. A generous private income fostered a fascination for fast cars and flying, and his interests were indulged financially by his wealthy uncle, Tom Hewitt, owner of the Tower Brewery in Grimsby.[4]

Hewitt at Harrow. (Courtesy the Hewitt Papers)

For centuries people had been attempting to take to the air when, in December 1903, the Wright brothers succeeded in making the first sustained flight in a powered, heavier-than-air machine. Their success marked the beginning of a new era in manned flight, and everyone was excited by the possibilities of this new technology. The *Daily Mail* newspaper offered £1,000 for anyone who could fly across the English Channel. The prize was won in July 1909 by the French inventor and aviator Louis Blériot in a contraption of his own design. Fuelled by his success, Blériot set up a research establishment and business, Blériot Aéronautique, that built and sold aircraft.

Enthralled by Blériot's achievement, the 21-year-old Vivian Hewitt started to experiment with large gliders, although his ambition was powered flight. The purchase of an Antoinette aeroplane – a gift from his wealthy uncle Tom in 1909 – saw the beginning of Hewitt's flight to fame. Towards the end of that year he rented a shed at Brooklands – the hub of British motor racing (and later its first airfield) – audaciously advertising himself as 'a dealer in second-hand cars and aeroplanes'. Surrounded by like-minded engineers, designers and aviation enthusiasts, Brooklands was exactly the environment Hewitt needed.

The Antoinette was a difficult machine to fly. A few days before Blériot's celebrated Channel crossing, another Frenchman, Hubert Latham, had attempted to do the same, but his Antoinette had ended up in the sea. The crux of the problem was that unless the air was perfectly still, the Antoinette was unmanageable. In 1910 Hewitt went back to Uncle Tom, who gave him the £1,100 – equivalent to £107,000 ($136,000) in today's money – for a Gnome-Blériot monoplane. Uncle Tom's generosity seemed to know no bounds. He clearly wanted his nephew to succeed, for later that year he gave Hewitt the funds for a second Blériot aeroplane.

21 West Kinmel Street, Rhyl, with Vivian Hewitt's car parked outside. This photograph dates from around 1910–1914. (Courtesy of Colin Jones, Rhyl History Society)

After Hewitt's father, Titus, died in 1910 of appendicitis at just 44, Hewitt's mother became increasingly eccentric and religious. Vivian, by now aged 22, found her difficult to deal with and decided to leave home. He moved to Rhyl in north Wales, where he took lodgings with the parents of John Parry, a man he had met while in Crewe. Parry's mother ran a boarding house, coincidentally named Holmfield, at 21 West Kinmel Street. When Hewitt became a lodger there in 1910, John Parry had recently married the vivacious and attractive Eleanor (Nellie) Myfanwy Jones. Three years later Parry's mother died, after which John and Nellie took over the boarding house with Hewitt still lodging there, and by now very much part of the family.

In October 1911 'Mr Hewitt surprised the residents of Rhyl by making two splendid flights over the town'. These flights were a complete novelty. Thrilled by their daredevil countryman, the residents came out in their droves to watch. The *Rhyl Pilot* proudly reported in April 1912 that 'Mr Hewitt has been most successful in his new "Blériot" and has flown nearly 6,000 miles in it.' Loving the attention, Hewitt had his name painted boldly on the underside of his wings and showered printed cards 'Dropped by Vivian Hewitt' onto the crowds below. He was flying high.[5]

His ambition was to beat Blériot's 20-mile flight by being the first to fly the 60 miles across the Irish Sea. Hewitt wasn't alone; there were others equally keen to make the Irish Sea crossing. Desperate not to be beaten, Hewitt was ready to go by 23 April 1912. But it was not to be. Thick mist rendered flying unsafe. For three long days Hewitt waited at Holyhead. Then, on the morning of Friday 26 April, the day dawned bright with a favourable wind. At 10.30 in the morning, watched by a huge crowd, Hewitt was off. Rising high into the air, he was soon surprised to encounter a bank of thick fog. There was also a breeze that blew him further south than he intended. Finally, emerging from the mist, he caught sight of Ireland's Wicklow Hills and was able to correct his position. Just 90 minutes after take-off, he landed safely in Dublin's Phoenix Park. 'All day long, crowds from Dublin and the surrounding country were coming … to see the beautiful white-winged machine … and the intrepid aviator – the first to cross the Irish Sea from Holyhead to Dublin.'[6]

The day after Hewitt's flight 'every newspaper, British and Irish, carried an account of the epic flight'. Hewitt was now a celebrity, and the people of Rhyl laid on a wonderful reception for his return. Despite all the fuss, Hewitt 'remained modest and unassuming'. He had been lucky, since other would-be aviators had died attempting similar feats. As a reporter writing for the *Observer* noted wryly: 'Aviators … really seem to possess less common-sense than their fellow creatures.' Although neither he nor Hewitt knew it, there was a certain

Vivian Hewitt and the plane in which he crossed the Irish Sea in 1912. (Hewitt Papers)

irony here, for that is exactly how people thought about the great auk. An airborne aviator and a flightless bird bound together in their supposed stupidity.

With the start of the First World War, Hewitt put his aviation and engineering expertise to good use by working with the military at Farnborough, helping to develop methods for firing machine guns from flying aircraft. He avoided being sent to the Western Front because of an apparent 'heart lesion' that rendered him unfit for active service. He must have been mightily relieved, for his much-loved younger brother, Billie, who had enlisted in the army, died soon after his arrival in France in 1914. Instead, Hewitt spent time in the United States acting as a test pilot, where he was promoted to Lieutenant in 1915 and to Captain in 1919. 'Captain' was how he was known subsequently to both his friends and family.

On one occasion in 1918, while still in the United States, Hewitt was forced to make an emergency landing in a ploughed field. On impact, his head hit the instrument panel and he was concussed. Concerned that this accident might result in blackouts, he was advised to stop flying. It must have been a bitter pill to swallow. He switched instead to high-powered motor launches and was soon busy testing these. By 1919, however, Hewitt had returned to the Parrys' boarding house in Rhyl, where his room had been kept ready for him throughout the war.

With funds once again provided by Uncle Tom, Hewitt bought his own boat, and in 1920 began visiting the Welsh seabird islands. Egg-collecting was widespread in Britain at this time and Hewitt was simply one among a vast army; this hobby gave his boating trips both a purpose and a thrill. He loved islands, and at one point even considered buying Bardsey, which lies off the tip of the Llŷn Peninsula and is arguably the most beautiful of all the Welsh Islands. The residents resented his presence, however, and were so rude to him when he was there that he changed his mind. His favourite place for his boating ventures was Puffin Island off the eastern tip of Anglesey. One year in the mid- to late 1920s he rented the island from the owners so he could watch and enjoy the puffins and other seabirds that bred there. Hewitt also rented

Eleanor (Nellie) Myfanwy Parry and Vivian Hewitt at Glan y Mor (Marble Quarry House), one of the properties in Penmon that they stayed in, around 1925. (Courtesy the Anglesey Archives)

properties in Penmon, a tiny cluster of houses on the Anglesey mainland opposite Puffin Island, where he was accompanied by his young 'housekeeper' Mrs Parry, who seems to have switched from 'landlady' to 'housekeeper' at around this time, and her children.[7] It is easy to see why Hewitt fell in love with Penmon: it is remote and beautiful, with views stretching out across the wave-washed Menai Straits into the distant mountains of Snowdonia on the mainland.

Other images of a happy Hewitt on one or other beautiful seabird islands make it all too easy to imagine how he became hooked on seabirds and their eggs. Like a virus replicating deep within him, his oological obsession had started to grow, but little could he have anticipated how much further it had to go.

For those egg-collectors more interested in building a collection than hunting down specimens for themselves, buying eggs from dealers was the easy option. Charles Harold Gowland was one such dealer, and someone who was to play a crucial role in feeding Hewitt's addiction. A keen egg-collector himself, Gowland was employed as a clerk by a successful Liverpool shipping company, H. E. Moss & Co. In February 1924 he came across an advert Hewitt had placed in a magazine called *The Oologists' Record*, expressing his interest in buying seabird eggs. Gowland wrote to him: 'I understand ... that you are anxious to complete a certain set of eggs' – a letter that launched many hundreds more over the next three decades.[8]

As well as finding nests and taking eggs himself, Gowland bought eggs from a network of collectors, making full of use of his position with Moss & Co. to import eggs from all over the world. He brokered deals between those forced to relinquish their collections through illness, death or debt, and

WANTED

Sets of Accipitres, with data, from all parts of the world, also rare types of British breeding sea birds. Will pay cash or exchange rare material. Have many sea birds' sets on the British, American, Indian and Australian Lists to exchange for Accipitres.

CAPT. VIVIAN HEWITT, HOLMFIELD, RHYL, NORTH WALES.

Hewitt's advert requesting the eggs of accipiters (hawks) and seabirds from 1924, indicating that by this date he already had eggs of other species he was prepared to exchange. (From *The Oologists' Record*)

those desperate to add to their collections. Effectively, he was the infundibulum – a vast, funnel-shaped conduit – through which all eggs passed between the birds that laid them and their human collectors.

Gowland's clients spanned the entire spectrum of collectors, from schoolboys able to afford only a single egg to wealthy, middle-aged men seeking something unusual. Schoolboys collected indiscriminately; adults were more discerning, acquiring the eggs of rare species, or unusual eggs of common species. Some collectors simply amassed huge numbers of eggs to illustrate the sheer variety of shape, colour and pattern. Butterfly and moth collectors were the same, desperate for 'varieties', and, as a result, they also often accumulated huge 'series' of specimens. Such variation is typical of all life forms, exemplified by the distinctiveness in human faces, voices and odours. The world would be a rather unexciting place without this variation, and without the variation in birds' eggs there would have been little or no urge for collectors to collect.

Variation in biological traits usually follows what is called a 'normal distribution', a bell-shaped curve depicting the

Harold Gowland aged about 34, in 1933. (Courtesy of Jim Whitaker)

relative numbers of different types. The eggs of the herring gull, for example, are typically olive or khaki, overlain with a smattering of darker blotches. The eggs laid by individual gulls may be paler or darker, browner or greener, and the spots may be larger or smaller. At the extremes of the distribution – that is, rarely – herring gull eggs may be sky blue, or even red with or without spots. In August 1927, Gowland wrote to Hewitt – who was collecting keenly by this date – telling him that he had just received a clutch of very rare *red* herring gull eggs from Norway. This was only the second such clutch he had seen in five years, he said. 'They are expensive because they are rare, but also because of the cost of importing them from Norway; the cost of translating etc.'[9]

For some oologists, aberrantly red eggs were a favourite. In that same letter, Gowland told Hewitt that he had also just received 400 guillemot and razorbill eggs from St Kilda. Among them, he said, is 'the finest Razorbill egg I have seen … Numerous people have asked me for these rarities, but I have remembered my previous promise, and you have first chance on them.'

We don't have Hewitt's reply, so we don't know whether he was tempted by Gowland's offer of the red gull eggs. I suspect he bought them, since – long after his death – I found several clutches of red gull eggs in Hewitt's collection.[10] Inevitably, Gowland was soon trying to tempt Hewitt with additional offers. Recognising that he was both keen and rich, Gowland courted Hewitt with predatory ardour.

The following year, in May 1928, Gowland informed Hewitt of another extraordinary collection of guillemot eggs that the owner was thinking of selling. 'If you would be interested, I will be only too pleased to make the necessary arrangements … I can safely say there is no other collection in the world which contains the variation, and can also safely say that there never will be another collection like it.' Gowland told him that for the six cabinets, containing well over a thousand eggs, Frederick George Lupton – whose collection it was – wouldn't accept anything less than £1,150 (equivalent to around £60,000/$76,000 today); it seems that either Hewitt declined or Lupton changed his mind, for it was not until several years had passed that Hewitt finally got his hands on Lupton's spectacular collection.

Then, just a month later, in June 1928, Gowland was back with yet another enticing offer. This time, it was Lupton's collection of lapwing eggs. Lupton was a solicitor based in Accrington in Lancashire. He was hopeless with money, however, and often in debt. These offers, made through Gowland, were probably the result of Lupton being short of funds. The Lupton family later told me that Lupton's attitude to money was both careless and odd: travelling by train between his home and Bridlington to reach Bempton to buy guillemot eggs, he prided himself on never buying a ticket.[11]

Lapwing eggs were once collected in almost unbelievable numbers, their contents consumed by wealthy clients (as 'plover's eggs') in posh London restaurants. Lupton said that between 1900 and 1924 he examined more than 100,000 lapwing clutches obtained by local collectors in the Settle region of north-west England. Among them was a *single* clutch of sky-blue eggs. Normally camouflaged to match the marshy grass background on which they are laid, lapwings, like gulls, very occasionally produce blue or sometimes red eggs. Once again, we do not know whether Hewitt bought Lupton's lapwing eggs in 1928, but he subsequently did acquire the most extraordinary collection of aberrantly coloured lapwing eggs.[12]

Gowland was a contradiction: an avid collector and dealer in birds' eggs, but at the same time interested in the conservation of birds. By the end of the Second World War, Gowland's oological business was sufficiently successful that it had become his main source of income. Spurred on by this, he started to publish a magazine, *Birdland*, in 1946, aimed at young naturalists. *Birdland* made a fabulously disingenuous disclaimer: 'Although many of the articles in this magazine will deal chiefly with egg-collecting, it cannot be too emphatically stated that we do not agree with, or encourage, the indiscriminate taking of eggs.' I spoke to a man called Derek Cotgrave, aged 91, who remembers seeing Gowland's adverts in boys' comics in the 1940s. He told me about the excitement he felt when the small wooden boxes of eggs he had ordered arrived in the post. Cotgrave once visited Gowland's home, where he was shown a stuffed great auk, and a great auk egg that he kept in a cabinet in his garden shed. Although genuinely keen to encourage youngsters, there's little doubt that Gowland's *Birdland* was a marketing vehicle for his egg business. Despite being reviled by bird protectionists, Gowland wasn't all bad. He was the first to attempt to save the red kite from extinction in Britain, importing kite eggs from Spain and fostering them under buzzards in the Welsh Valleys. That attempt was unsuccessful, but it undoubtedly inspired a later effort in 1989 that *was* successful in securing the UK's kite population.[13]

People collect all sorts of things, from matchboxes to postage stamps, porcelain and Pokémon cards. How should we think of them? Part of the way we regard collectors depends on whether what they collect is legal to own or not. Collecting books is legal, collecting birds' eggs generally is not, at least not now. Book collectors are harmless because their collecting may help authors. Egg-collecting is not, for two main reasons. First, because taking eggs is now illegal. Second, because serious egg-collectors usually focused on birds whose populations were small and vulnerable, like British ospreys and red-backed shrikes. There is also a third element: taking and blowing eggs involved depriving their embryos of life, and as such can be considered immoral.

Addictive hoarding is a recognised medical or psychological condition. Addictive collecting, however, is not. If it were to be recognised as such, would we think differently about egg-collectors? It seems analogous to other addictions. Drug addicts take stuff (sometimes from our homes) illegally to feed their habit, and we despise anyone who steals our belongings. But there are some who recognise drug-induced addiction for what it is and, while not condoning burglaries, advocate psychological or medical help. Egg-collecting is stealing from nature – that is, from all of us. There is almost no sympathy for egg-collectors today, especially if they are working class; wealthy egg-collectors, like wealthy drug addicts or wealthy drug dealers, are no less despicable but are much less likely to be apprehended. This may be one reason why Hewitt avoided prosecution, despite the large, yolky, bloody, posthumous stain left by his collections.

Hewitt's uncle Tom died in 1930. As his sole heir, Hewitt inherited a fortune, as well as 'many large estates' including parts of Grimsby Docks, businesses and commercial property, and 300 tied houses. Already well off, Vivian Hewitt was now staggeringly wealthy, with an annual income of around £50,000 – equivalent to a breath-taking £2.7 million ($3.4 million) today. His earlier propensity to collect birds' eggs and other items now burst like an exuberant flight of racing pigeons from a loft. Financially unrestrained, his collecting frenzy expanded to include guns, coins, jewellery, stuffed birds and more eggs on an almost unprecedented scale. He was now able to purchase entire collections accumulated by others, either directly or through Harold Gowland.

In the year he inherited his fortune, Hewitt was 41. Mrs Parry was still his housekeeper and, together with her eldest son, Jack, he started to search for a home for them all. North Wales was where Hewitt wanted to live and, after looking for likely properties along Anglesey's northern coastline, he and Jack eventually arrived at the enormous shingle beach of Cemlyn Bay. It was love at first sight. Hewitt wanted somewhere isolated and surrounded by birdlife, and Cemlyn seemed perfect. The house, known as Bryn Aber, was hardly luxurious, for it had neither electricity nor running water, but that aside, it was everything Hewitt wanted, and was to be his home for the next 30 years.

Hewitt, Mrs Parry and three of her children – Jack, Myfanwy and Vivian – moved into Bryn Aber in 1931. Mrs Parry's other son, Ken, had left home when he joined the army as a teenager. In 1931 Jack was 21, Myfanwy – always known as 'Girlie' – was 18, and Vivian was 16.[14]

After inheriting his fortune, Hewitt became paranoid about taxes and officialdom, so much so that he put Bryn Aber in Nellie's name. As if the house wasn't secluded enough, soon after moving in Hewitt employed a team of workmen to build a huge, 10-foot-high wall around the entire property. It was, he said, to provide a place of sanctuary for the birds by protecting the trees from salt spray. But it was a sanctuary for him, too, and effectively another Welsh island, like those he loved to visit in search of eggs. The wall, whose construction went on for years, provided employment for several local men. When I first visited Bryn Aber in 2022, my immediate impression was of a lonely, impregnable fortress. Hewitt was a man who wished to remain largely hidden from the world. Apart from William Hywel Jones, Hewitt's doctor, hardly anyone was allowed inside the house. To further their isolation, Hewitt eventually bought much of the surrounding farmland.

Hewitt's wall-builders of Bryn Aber. (Hewitt Papers)

Mrs Parry (far right) with her four children. From left to right: Viv, Girlie, Ken and Jack. This photo was taken around 1941. (Courtesy the Anglesey Archives)

Bryn Aber was isolated, and the living conditions primitive, yet Mrs Parry and the children took it all in their stride. As did Hewitt. But when he went up to London – travelling by train from Holyhead – he stayed in one of the capital's most expensive and luxurious hotels, the Savoy. He was a shareholder there, and on each visit, Hewitt was given the same seventh-floor suite, cared for by the same butlers. When Mrs Parry accompanied him, they occupied separate rooms.[15]

Hewitt slipped effortlessly into Bryn Aber life, cared for – loved even – by Nellie, and surrounded by the infectious enthusiasm of the three children. Family photographs show how close the siblings were, both to each other but also to Hewitt. They radiate happiness, and in some respects so they should have, for with Hewitt's enormous wealth they probably wanted for little (other than basic home comforts like running water and electricity). They had fast cars and motorcycles, and, like him, clearly loved Cemlyn's remoteness and beauty.

The locals thought Hewitt very odd, but he was no snob and was as comfortable talking to his workmen as he was to those as rich as himself. To those who rented properties on his land, Hewitt was extremely generous, especially at Christmas. At home he was unconcerned by his appearance, wearing comfortable old clothes and allowing his hair to grow long. Yet, when visiting London, he appeared well groomed, with smart suits and cigars. Mrs Parry, attractive, gracious and full of fun, had the patience of a saint. She smoothed and organised Hewitt's life, accommodating his demanding and often peculiarly rigid ways. That Mrs Parry was devoted to Hewitt there is no doubt, for she cared for him with unbounded patience and tact. 'A determined and rather self-centred man', Hewitt could be extremely demanding, bellowing 'Nellie!' whenever he needed something. The house was filled to the brim with Hewitt's endless purchases: books, stuffed birds and cabinets of eggs. One of the most bizarre aspects of Hewitt's obsessive collecting was that once a parcel had been delivered and he had opened and inspected its contents, he simply repacked it as precisely as possible. He probably never looked at it again. Acquisition was everything.[16]

Hewitt first met Nellie through his friendship with John Parry in 1908, when she was Parry's fiancée. Understanding Hewitt and Nellie's friendship is crucial if we are to ever unravel the rumours regarding their relationship and the paternity of her children. When they first met, Hewitt was 20 and she was 25. They were both glamorous in their different ways, and it is not difficult to imagine an immediate mutual attraction. Exactly when John Parry and his wife separated is unknown since Hewitt's biographer, Hywel Jones (actually Hewitt's doctor, William Hywel Jones, writing under a pseudonym), is extremely circumspect on this issue. When Mrs Parry eventually moved out with her children to become Hewitt's 'housekeeper' and live-in companion at Penmon in the early 1920s, she was in her early forties and Hewitt was 37. We know little of John Parry other than that he continued to live in his mother's house in West Kinmel Street, Rhyl, while his wife and three of his children lived with Hewitt. Significantly, perhaps, Hewitt bought that house at some point, raising the question of whether Parry might have been simply bought off by Hewitt. John Parry died in 1948, some 23 years after Nellie had set up home with Hewitt.[17]

The writer Glen Chilton, pursuing details of Vivian Hewitt's purchase of a stuffed specimen of another extinct bird, the Labrador duck, says that Mrs Parry's children referred to Hewitt as 'Dad' around the house, but 'Captain' elsewhere.[18] Attitudes to illegitimacy and co-habiting in the 1930s were very different from those of today. If any of the children were Hewitt's biological offspring – and he treated them all as though they were – neither he nor Mrs Parry was ever going to tell anyone. Nor are there any written records. When I examined an archive of family photographs, I felt I could see a certain Hewitt likeness, certainly in two of her children, Vivian and Girlie.

That Hewitt and Nellie were close is indisputable. Hywel Jones describes them as being 'like brother and sister', which could be true, but he may merely have been protecting their honour. There is a photograph of the couple boarding a Pan American flight in New York in 1962, with a press caption that states: 'Captain Vivian D. Hewitt and his sister Mrs E. M. Parry.'

I wonder whether *sister* was what Hewitt told the photographer? There's another rather candid photograph of the couple arm in arm, taken outside an expensive restaurant in Dublin, suggestive of a relationship closer than employer and housekeeper. Also, was it a coincidence that Mrs Parry's fourth child was named 'Vivian' – an unusual name at the time? And then there's Hewitt's extensive correspondence with Mrs Parry's eldest son, Jack, in which Hewitt's affection is so apparent. Hewitt addressed him each time as 'Dearest Dumbo' and signed off: 'love, Captain' followed by several kisses. One letter, from 9 May 1963, finishes: 'You mean a very great deal to me, my Elephant'.

As the years went by, Harold Gowland continued to court Hewitt, offering ever more tantalising collectables.

(Left) Vivian Hewitt and his 'sister', Mrs Parry, about to board the PanAm flight in New York that carried the airline's millionth passenger in 1962; (Right) Outside the Metropole Hotel, O'Connell Street, Dublin, arm in arm, possibly on the same trip. Probably taken by the street photographer Arthur Fields, known as 'the man who stood on O'Connell Bridge taking pictures'. (Hewitt Papers)

Understanding the collector's mind better than any psychologist, Gowland played Hewitt as though he was fishing; laying out a tempting bait at just the right time, he waited for the bite, and, once 'on', took time to drive home the hook before reeling in the prize. Gowland knew that Hewitt was the biggest fish in the oological pond – the ultimate specimen – that, if caught, could be released and re-caught time after time. He was the ardent suitor, courting, cajoling and fawning with a persistence few others could match. Gowland had Hewitt's measure, and his letters are carefully concocted messages, full of promise.

Poised to strike, and to make a substantial profit, Gowland wrote to Hewitt in April 1933:

> *Seeing you are interested in eggs of the sea birds of England, what about a Great Auk egg? I have one for sale …*

Hewitt had previously bought plenty of eggs from Gowland, but, prior to receiving this letter, he had probably never thought about purchasing and owning a great auk egg – the ultimate oological prize. For Hewitt, once that particular nuclear button had been pressed there was no going back.[19]

Chapter 8

Spend, Spend, Spend

We do not know whether Hewitt was tempted by Gowland's great auk egg offer. Nor do we know which egg it was or who it had belonged to. It didn't matter, for less than two years later a tremendous haul of six great auk eggs and two skins came onto the market. Once the property of George Dawson Rowley (who we met earlier – see p. 87), these specimens had passed to his son Fydell after his death in 1878, along with all of Rowley's many other eggs and skins. After Fydell's death in October 1933, the family decided to put the entire collection up for auction at Stevens's Auction Rooms Ltd in Covent Garden.

Stevens's had a long history of selling birds' eggs and skins, but this was a sale like no other. Collectors were drawn from all over the world and, as one observer explained, it was the great auk specimens they had come for. Six eggs together was unprecedented. The largest number offered previously was four on 11 July 1865, also by Stevens's.[1] Indeed, Rowley's personal collection of great auk eggs was trumped only by the seven in Cambridge University Museum, by a Robert Champley of Scarborough who had nine, and by the ten owned by the Royal College of Surgeons (that they were unaware of until the eggs were discovered there by Alfred Newton in 1861).[2]

The sale of *two* stuffed great auk specimens – both in summer plumage, and said to be a male and a female – was unusual too. The alleged female had been bought in 1869 by Rowley from the dealer Gustav Frank (son of another great auk dealer, Johann Heinrich Frank, who supplied auk material to Naumann). Frank had originally acquired it, and another one, in 1845–46 from the dealer known as 'Israel of

Dawson Rowley's two stuffed auks and six eggs, with a huge elephant bird egg. Photographed in 1934 at the time of the Stevens's sale. (From Fuller, 1999)

Copenhagen'. Professor Japetus Steenstrup, who was once dubbed the 'father of great auk history', thought these two birds might have been the very last two known, the pair killed on Eldey in 1844. Stevens's Auction Rooms was obviously unaware of this possibility, for, had they known, they would no doubt have used it to ramp up the auction price. Nor were any of the great auk hopefuls present at auction – including Vivian Hewitt – aware of this tantalising possibility.[3]

Bidding started at two hundred guineas, and ended when Hewitt offered five hundred – an extraordinary sum, equivalent to £31,000 ($39,000) today – for the bird now known as the Los Angeles Auk. The other specimen – the alleged male – had been acquired by Rowley in 1868 from James Gardner, a London taxidermist who had bought it in 1848 from a dealer in Paris. The bidding for this one also started at two hundred guineas, and was secured by Hewitt again for five hundred.[4] This was the bird now known as the Cincinnati Auk.

Vivian Hewitt and the two Rowley great auks he purchased in 1934. (From the *Daily Sketch*, via David Clugston)

As far as the six eggs were concerned, Hewitt showed uncharacteristic restraint and bought only two for a total of three hundred guineas (£18,500/$23,500 at the time of writing). Using their modern names, these were Dawson Rowley's Egg and Lord Garvagh's Egg. They were the most beautiful and the most expensive of the six eggs sold that day. It is understandable why Hewitt turned down two of the other eggs, for they were damaged, but it isn't clear why he allowed the remaining two to slip through his fingers.

It was an expensive day out, for the two eggs and two skins together cost Hewitt £1,659 – around £98,000 ($125,000) at today's prices. Harold Gowland, who had also been at the sale, wrote afterwards to Hewitt, saying 'I hope you arrived home with your treasures quite safe. I told Mrs Gowland all about it on my arrival home last night and she said she wished she had been there to see the excitement.' On the day after the sale, the *Daily Sketch* newspaper carried a picture of Hewitt, cigar in hand, looking admiringly at the two stuffed specimens he had secured.[5] That he was pleased with his purchases is an understatement. When the two auks duly arrived at Bryn Aber – via the local postman – Hewitt could not contain his excitement. Summoning the men who were busy building the wall, Hewitt launched 'into an oration on the Great Auk, its habits and history, finally concluding with the price he had paid'. Given that Hewitt's workmen were probably earning about £1 a week, they were understandably stunned. One of them spluttered: 'Pay me that much and I'll go in a ruddy cage myself and sing to you.' For someone generally so non-elitist, Hewitt's oration was extraordinarily insensitive.[6]

A year later, Hewitt's doctor, William Hywel Jones, completely unaware of Hewitt's purchases at the Stevens's sale or the incident with the builders, was chatting with Hewitt in his bed-sitting room. Still hugely excited, Hewitt opened a drawer to reveal, hidden amongst his underwear, a box containing the two great auk eggs. With no interest in birds or their eggs, Jones looked on with amazement. 'It's an investment,' said Hewitt. 'Every year they will increase in value.' Like the builders, Jones could not believe the cost, and was left wondering what he might have done with that amount of money.[7]

On 1 July 1935, Gowland wrote to Hewitt to say that he had seen F. G. Lupton (see p. 92) in London:

It was very fortunate that I did, as on account of various circumstances over which he has no control, he is obliged to sell his guillemot [egg] collection. As you are aware, he is a widower, and for the past few years has been in lodgings in London ... he is not a methodical man at the best of times ... his various landladies get fed up with having their rooms filled with what they term rubbish ... He has the pick of practically every egg which comes up from Bempton, and has purchased every well-known guillemot collection that has come on the market ... he has them arranged in a most wonderful and scientific way.

Gowland finished his typed letter with a hand-written flourish: 'If you were to purchase these eggs, it would represent, in my honest opinion, the most important purchase you have ever made – even including your great auk purchase.'[8]

Back in 1928, Gowland had offered Hewitt Lupton's guillemot and lapwing eggs, but one or other of the parties changed their mind before a sale could be clinched for either collection. This time, Hewitt was determined not to miss out, especially as Lupton's collection included another great auk egg. This one – Lady Cust's Egg (see p. 96) – like many others, had a convoluted pedigree. It was one of the six once owned by Dawson Rowley that, at the sale of 1934, had been bought by the Reverend Francis Jourdain, who later swapped it with Lupton for another, known as Yarrell's Egg.[9]

Oological enthusiasts. From left to right: F. G. Lupton, the Reverend Francis Jourdain and Vivian Hewitt. (Courtesy Verity Peterson and the Hewitt Papers)

For the single great auk egg and all the guillemot, razorbill and puffin eggs, Hewitt paid Lupton £1,600 (equivalent to £94,000/$120,000 at the time of writing). The receipt, dated 29 July 1935, shows that Hewitt paid a further £181 for the eggs of a variety of other species, mainly birds of prey, that he was especially interested in.

Also listed on that receipt was £20 that Hewitt paid Lupton for some model great auk eggs. Replicas of great auk eggs were not uncommon. Usually made of plaster of Paris and cast from a mould, they were then painted by hand, sometimes to mimic specific eggs, sometimes in a more generic and less authentic way. The models that Hewitt bought from Lupton, though, were exquisite. Turned in light wood and painted with photographic realism, they looked almost as good as the real thing.[10]

In an extraordinary piece of good fortune, I was given four of those replica eggs in 2021. They included one that was a copy of an egg named 'Dr Dick's Egg' that was one of the ten great auk eggs discovered by Alfred Newton in the Hunterian Museum. This egg is especially beautiful, with a fawn ground colour, overlain with fine, dark pencil-squiggles. Years before, I had imprinted on this particular egg – much as the parent great auk that laid it would have done (see p. 99). During my first year as a Zoology lecturer at the University of Sheffield in 1976, there was a ramshackle second-hand bookshop – now long since demolished – opposite the university. One day, in the window, I saw a copy of Henry Seebohm's *Coloured Figures of the Eggs of British Birds*, published in 1896. Seebohm had been a Sheffield steel magnate, and his book was famous for its beautiful plates, but I never imagined I might own a copy. I bought the book and was thrilled to discover that it contained a plate depicting this great auk egg. Seebohm writes that *he* was allowed to copy the egg in the Hunterian Museum, but in truth it was an unacknowledged artist who copied the egg.[11]

The Hunterian Museum was based on the collections that once belonged to the renowned anatomist and surgeon John Hunter in the 1700s. Newton noticed that two of the ten eggs were in a box labelled 'Penguin eggs from Dr Dick' but he failed to discover the identity of the doctor.[12] The name 'penguin' on the box provided a pointer, in as much as it suggested a French, and hence a Newfoundland, connection. There were no further clues until the 1990s when seabird biologist and great auk enthusiast Bill Bourne discovered the existence of a Robert Dick, a ship's surgeon mentioned in the UK's Royal Navy List, and 'active' between 1809 and 1844. I found the same Dr Dick in another list of 1809, where he was described as 'unfit'.[13]

A plausible but unverified explanation for John Hunter's 10 eggs is that they were a gift from Captain George Cartwright. An ex-army officer, Cartwright spent almost 20 years between 1766 and 1786 wheeling and dealing in furs and politics on the coasts of Newfoundland and Labrador. The settlement he established in Sandwich Bay now bears his name. Cartwright's journal, published in 1792, provides an extraordinary insight into both his life as an adventurer and the country's indigenous people and natural history, including the great auk. Cartwright commented on the 'singular and almost incredible fact that these people [the Beothuk] should visit Funk Island … and repair thither once or twice every year and return with their canoes laden with birds and eggs; for the number of sea-fowl which resort to this island to breed are far beyond credibility'.[14]

In 1773, during one of his regular visits to England, Cartwright was invited to dine with John Hunter at his home. I can readily imagine him arriving with a gift of great auk

Cartwright's 'indians', Labrador Inuit. From left to right: Tuglavingaaq, his wife Qavvik, Atajug (Tuglavingaaq's brother) and his wife Ikkanguaq, with their young daughter, Ikiunaq, just behind. A pencil drawing by an unknown artist. (Courtesy of the Hunterian Museum, Royal College of Surgeons)

eggs as a contribution to Hunter's rapidly expanding natural history collection. However, the main motivation for the dinner invitation was so that Hunter could see the five Inuit people Cartwright had brought to England from Labrador. Now deemed a reprehensible act of colonialism, it was not uncommon for those travelling to remote regions of the world to return with human souvenirs.[15]

Cartwright's 'noble savages' – Atajug, his wife Ikkanguaq and their young daughter Ikiunaq, plus Atajug's brother Tuglavingaaq and his wife Qavvik – caused a sensation in London society. They were viewed enthusiastically by King George III, Samuel Johnson, James Boswell and Joseph Banks. Banks had visited Newfoundland himself in 1766 and later became president of Britain's Royal Society.[16]

As well as being gawped at by John Hunter and other curious Londoners, Cartwright's 'indians' were said to have enjoyed the glories of the Georgian capital, perhaps in the same way one might savour being dumped onto another planet.

Keen to ensure that they saw the best of London – to impress and improve his charges – Cartwright arranged for them to see a performance of Shakespeare's *Cymbeline*, and to visit a shop in Piccadilly that sold animals, both of which left them mystified. On seeing a monkey in the shop they asked 'Is that an Eskimo?' In a sad and all-too-familiar story, when setting off back to Labrador in May 1773, they had not even left the Thames estuary when Qavvik became sick with smallpox. Inevitably, her four relatives soon contracted the disease as well; all of them died before leaving England's shores. Remarkably, Qavvik herself survived, only to become on reaching home what we would now call a 'super-spreader'. Thanks to Captain George Cartwright, she was unwittingly responsible for exterminating almost the entire indigenous population of southern Labrador.[17]

Dr Dick remains unidentified. Notwithstanding the ship's surgeon found by Bill Bourne and me, the likeliest candidate is perhaps Dr Elisha Cullen Dick, a physician and politician living in Philadelphia, who studied anatomy and was a close friend of John Hunter. Elisha Dick may have given Cartwright the gift of great auk eggs to pass on to Hunter. Equally mysterious is how Vivian Hewitt acquired this and two other eggs in 1946 that were also part of the ten Alfred Newton had found in the Hunterian Museum all those years before.[18]

Throughout the 1930s and early 1940s, Hewitt continued to snap up the egg collections of other oologists as they became available. These included the collections of the Rev. Jourdain and of Edward Stuart-Baker, both of whom were respected scientific ornithologists. Interestingly, as these two individuals exemplify, the opprobrium of oology could be offset by science – something that Hewitt lacked. Jourdain was an outstanding all-rounder and a crucial contributor to the *Handbook of British Birds*. Baker was more of a specialist, best known for his studies of cuckoos and other brood parasites. The exact dates and circumstances surrounding these acquisitions by Hewitt have not been recorded. This was, however, a period when many collectors found themselves hard up as a result of the Great Depression of the 1930s. In a letter to Peter Adolph, Hewitt recounted how 'during the slump in America I purchased all the good material which I could lay my hands on … for seven years prior to the War, I had a man in America who did nothing but collect American Accipitres [hawks] for me, not only taking the eggs himself but purchasing outstanding clutches from various well-known collections.' Hewitt's acquisition of other oologists' collections during the Depression was analogous to the acquisition of entire libraries by wealthy American collectors following the break-up of English aristocratic homes 20 years earlier.[19]

Not only was Hewitt buying collections of eggs, he was also acquiring entire cabinets of study skins and bulky glass cases of stuffed birds. They came from a variety of sources, including dealers like Rowland Ward, William F. H. Rosenberg, Edward Gerrard & Sons, and, of course, from Harold Gowland. Writing to Hewitt in 1935, Gowland tried to tempt him with 523 cases of stuffed birds owned by F. G. Lupton. Later, in 1941, Hewitt bought 224 skins from Gowland. Hewitt also acquired some specimens himself, through birds either shot or found dead, including his beloved pet parrots, all of which he packed off to Rowland Ward for professional mounting and preservation. To accommodate all his cases of stuffed birds, Hewitt simply added yet more sheds and outhouses to the spacious grounds within Bryn Aber's high walls.[20]

Hewitt continued to search for additions to the two great auk eggs bought at the Stevens's sale in 1934, and Lady Cust's Egg that came as part of the Lupton collection the year after. Over the next 10 years, he was able to secure just about every great auk egg that came on the market. One he bought in 1937 became the eponymous Vivian Hewitt Egg. Then there were two others obtained in 1939 following the death of Herbert Massey, another avid collector.[21] At the outbreak of the Second World War, Hewitt decided to move his belongings, including many egg collections, from the house in Rhyl (which Hewitt had bought, though it seems John Parry was still living there) to Bryn Aber, on the assumption that the Germans were unlikely to bomb Anglesey. However, when Hewitt purchased Jourdain's huge collection after his death in 1940, he had it sent to Rhyl, presumably because there was insufficient space at Bryn Aber. Jourdain's collection included two great auk eggs: Bowman Labrey's Egg and Yarrell's Egg, neither of which Hewitt would have dared to leave in Rhyl. In 1946 Hewitt somehow managed to secure three eggs from the Hunterian Museum. The first two were Dr Dick's Egg, and – bearing in mind that these names were only allocated in 1999 – Alfred Newton's Egg. The third was named after someone who, as we will see, was to become a central player in the culmination of this story – Jack Gibson.[22]

The 13th and final great auk egg that Hewitt obtained was the one that is now called Malcolm's Egg, secured, together with a stuffed great auk, in 1948. Apart from the two eggs Hewitt obtained at the Stevens's sale of 1934, the amount he paid for these prizes is unclear, partly because, when the eggs came as part of someone's collection, the great auk eggs were not priced separately, and partly because – as in the case of the three Royal College of Surgeons eggs – Hewitt never disclosed what he had paid for them. If all 13 eggs cost the same as the

Stevens's ones – three hundred guineas – then in total he spent about £4,000, or in today's currency around £235,000 ($300,000).

Hewitt's collecting seems to have been a protracted frenzy of acquisition. It was fuelled by Harold Gowland, and facilitated by the Depression. But Hewitt was far from indiscriminate in his collecting. To be worth having, the eggs in particular had to be accompanied by good-quality data, including where and when they had been collected.

Yet it *was* a frenzy, and one wonders what it was about Hewitt that drove him so relentlessly. One obvious answer is simply that he could afford it. He had the money to purchase whatever he wanted, and at auction he could outbid just about anyone else. But what were the underlying psychological drivers that turned Hewitt into such an obsessive, almost compulsive collector?

In her poem 'Gathering', Nina Bagley writes perceptively and empathetically about those who, while out walking, pick up things such as sticks, stones and the eggshells of wild birds to take home. Bagley's gathering is one end – innocent and opportunistic – of the collecting continuum. At the other is the pathological 'gathering' that epitomised Hewitt and others like him.[23]

The unabated quest for objects of desire shares some similarities with love and the overwhelming need to acquire a partner. This is exemplified by Gabriel Garcia Marquez's character Florentino Ariza, in *Love in the Time of Cholera*, who waits for years to 'acquire' his heart's desire, Fermina Daza. Unlike collecting, however, love is usually satisfied by success and the obsession expiated. Collecting knows no such release: the acquisition of a desired object brings only momentary relief – within minutes or hours, the urge emerges again, and so it goes on. As the American journalist Barbara Grizzuti Harrison observed, 'Collecting is like sex; satisfaction renews and creates new appetites.'[24]

At first sight collecting seems to be the same as hoarding, a recognised psychological condition that falls under the umbrella term of 'obsessive-compulsive disorder'. But hoarding, which was so obvious and so bewildering to Hewitt's doctor when visiting Bryn Aber, is the *consequence* of collecting, rather than its main cause. To identify the causes of compulsive collecting, we must look deeper than either hoarding or love; deeper, and further back in time. A common thread among obsessive collectors is that many of them suffered some kind of childhood trauma, such as the loss of a parent or a loss of parental affection. One result of this can be a constant need for reassurance, fulfilled by acquisition. Unrelenting collecting seems to be a compensatory activity for early life issues, bestowing a sense of control. Objects, whether they are birds' eggs or antique jades, don't make demands. A collection is where the collector rules, or at least retains a fantasy of control. I have spoken to a number of obsessive egg-collectors, and the early loss of a parent is a recurrent theme. Vivian Hewitt's father died in 1910 when Hewitt was 22, and his much-loved younger brother, Billie, was killed in France during the early days of the First World War.[25] Hoarders can be identified from brain scans, yet so far as I know, no one has compared the brain activity of collectors and non-collectors. If I were to do this, I'd present my two groups with the eggs of rare birds – maybe those of a great auk – see which regions of their brain light up, and check whether it is the same as in hoarders or love-obsessed individuals.[26]

Trying to understand Hewitt's extraordinary behaviour, I have been struck by the similarities with the now-discredited American psychiatrist, Arthur Sackler. After making a fortune from marketing prescription drugs in the 1940s, Sackler became an obsessive collector of ancient Chinese art. His colleague Paul Singer said this about him: 'Really serious collecting was driven by a pattern of arousal and release that was downright erotic ... the pulse beats faster, the beholder sees the beauty that he [*sic*] wants to own.' Sackler's wife also noted that it was the 'hunt' that excited Arthur: 'identifying some precious artifact and then figuring out how to claim it was a secretive, sensual process'. As Sackler's biographer states, 'new boxes would arrive at the Long Island house, full of exquisite objects ... The unboxing took on the spiritual aspect of a séance as Arthur lifted out the ritual bronzes ... as if communing with ghosts, touching history.'

> *Inside the house, the boxes were piling up ... Arthur was now purchasing Chinese art at such a clip that new acquisitions were arriving more quickly than the family could open them. Upstairs, downstairs, in the attic: there were boxes everywhere ... before long, the sheer volume of material that he owned had reached a point where it could not really be understood or kept track of by the human eye ... and still, Arthur did not stop. He collected relentlessly and insatiably.*[27]

Vivian Hewitt was the same:

> *He was probably overwhelmed by the amount he had collected. Boxes were stacked up high in various corners and, wherever there was a spare resting place not already occupied by model engines, stuffed birds in glass cases were in competition ... the landing, yet again, was tightly crammed with even more egg cabinets, mounted birds and stack upon stack of papers, magazines and other sundry items. For the family to reach*

their bedrooms, with only a candle to light their way, called for
a great deal of local knowledge and a certain amenity of
expertise. The haphazard and disorderly manner in which the
Captain stored his collection was, to say the least,
incomprehensible ... There seemed to be no attempt at
classification: indeed, after his death many boxes were found
still unopened and uninspected. His bed-sitting room, as well
as the passages, were stacked high with countless boxes, and all
available drawers were equally crammed.[28]

Hewitt's joy at the acquisition of his first two great auk eggs
and skins apparently elicited the same kind of electric charge
and euphoria that Sackler felt when acquiring an antique
jade. I'm sure, too, that for Hewitt, there was an element of
both sensuality and communing with ghosts in handling his
great auk eggs. I have experienced the same sensory
engagement on being allowed to hold one. Sackler eventually
gave his collections to the Smithsonian in Washington DC.
It was an act that brought him 'public recognition in a way
that advertising had not ... it offered the possibility of
immortality'. Hewitt, too, by amassing more eggs and bird
skins than anyone else, *may* have hankered after more public
recognition than he received from his 1912 Irish Sea flight,
which was rather eclipsed by news of the loss of the *Titanic*
11 days earlier.

The locals of Cemlyn Bay and the surrounding area considered
Hewitt a recluse and an eccentric. The interminable building
of Bryn Aber's great wall would have been enough. But there
was more. Hewitt's casual dress, and what his biographer
called his 'lax coiffure' while at home, once saw him mistaken
for a tramp – by a genuine tramp. He was, however, obsessed

by dental hygiene, cleaning his teeth four or more times each day and always, *always* with the same brand of toothpaste. Who else but an eccentric would live in, and subject his adopted family to, a home with no running water, no electricity, no telephone, and an outside WC? Who would spurn the convenience of a telephone only to conduct all their urgent business by telegram? Once Hewitt had decided to purchase something, business was *always* urgent. Impulsive and impatient to secure a deal, there were sometimes as many as 10 telegrams in a day, with the telegram delivery boy rewarded with a half-crown tip each time. Other visitors to the house were less welcome, and were usually told that Hewitt was not at home.

His daily routine at Bryn Aber was a rigid one that everyone else had to accommodate, for Hewitt's 'greatest joy was being master of his own time'. The day started with breakfast in bed at around 11, after which he pottered around with his engines and eggs until Mrs Parry would announce lunch with a 'long piercing blast on a police whistle'. In the afternoon Hewitt often took a nap, after which he would spend an hour or so with his beloved pet parrots. He enjoyed and appreciated Nellie's cooking, and after dinner he continued to work late into the night. Unlike his uncle Tom, who was a connoisseur of fine wines, Hewitt had little interest in alcohol.[29]

Arthur Sackler was hugely egotistical and craved the appreciation of posterity. By contrast, Hewitt was remarkably humble and, while not craving immortality, like all collectors, he did want his collections to be preserved and appreciated after his death. Anticipating the outbreak of war two years later, Hewitt wrote his will in 1937. Its focus, apart from making provision for Mrs Parry and the children, is 'my eggs, bird skins, cabinets and bird books … My Trustees shall retain a sum of thirty-five thousand pounds [equivalent to £1.9m/$2.4m today] which shall be invested in some Trust Securities, the income so derived to be used in forming and keeping up a Museum for the Scientific study of Birds, Eggs and nests and my own collection shall find a resting place herein. It is be known as the "Vivian Hewitt Avian Museum".'

Hewitt's idea was that some of the funds would be used to acquire an old country house 'not less than one hundred miles from London' to avoid air-raids. He was adamant that no member of the British Ornithologists' Union (BOU) or any other organisation 'must ever have or be allowed to have control of this museum, as certain people are "Protectionists" and would do their best to put obstacles in the way rather than help the museum advance'. Similarly, he did not want any dealer in eggs to be part of the museum. 'The object of the museum', he said, was 'to build up large series of eggs of every form of bird in the world, taking into consideration fine preparation together with full and authentic data'. Finally, he stipulated that 'my great auk eggs and also the mounted skins of the great auk shall never be sold or exchanged, or leave the museum'.[30]

Hewitt shared his idea of a National Oological Centre with Francis Jourdain. The idea was that their two enormous, elite collections would form the foundation and that, in due course, others would add their collections to it. In an undated letter, Jourdain wrote to Hewitt, saying:

I wish we could meet to have a talk about your proposed museum etc. I am an old man now and life is such an uncertain business that I ought to put in writing what I am going to do … If your museum is to be a natural one, I feel very much inclined to give my collection to it. In many ways it is of course unique …

After explaining why this is the case, Jourdain continued:

> *There is also the possibility of my library going to the same place*
> *and I might help to endow it ... It could be made to take the place*
> *of Tring* [i.e. the Natural History Museum] *in the Bird World,*
> *especially if you included a skin collection as well ... Can you*
> *meet me in London or at home and see if we can settle anything?*

Whether that meeting ever occurred we do not know.
Jourdain died in February 1940, after which Hewitt's plans
seemed to falter.

In 1943 Hewitt wrote to Harry Witherby, a publisher and
one of Britain's best-known ornithologists: 'As things are at
present my own vast collection will after my time be left in
Trust for the benefit of ornithologists together with a sum of
£50,000 to provide for the upkeep of the same'. Then,
writing to fellow collector (and cuckoo aficionado) Edgar
Chance in 1946, Hewitt mentioned his plan to leave 'my
collection in Trust with a large sum but it may not be carried
out because of crippling taxation etc.' Hewitt repeated that
he did not want his eggs to go to a museum. In correspondence
with yet another collector, Francis Pitman, in March 1947,
in the immediate aftermath of the hardest winter in living
memory, Hewitt wrote: 'I wouldn't leave eggs of mine to any
museum ... None of them seem to take any interest ...
I cannot see a valuable and historical collection such as
mine treated in this way.'[31] Hewitt was right to be wary of
his eggs and skins going to a museum, but, sadly, the plan for
an oological centre came to nothing.

Mounted great auks were more difficult to acquire than eggs.
People tended to hang on to them, perhaps because they were
both more valuable and had more appeal. In 1936, Hewitt was

able to buy a third stuffed great auk, now known as the Clungunford Auk, to add to the two he bought at the Stevens's auction in 1934. He purchased it for £700 from the London taxidermy company Rowland Ward. Almost certainly from Iceland, this bird – or rather its skin – was owned initially by Gustav Frank in 1835, who eventually sold it to the ornithologist John Gould, who in turn sold it to Mr John Rocke, the owner of Clungunford House in Shropshire. Rocke had a huge collection of stuffed birds in his home and his great auk took centre stage in a wonderful diorama of North Atlantic seabirds. Rocke died in 1881, and some time after this the auk was bought by Rowland Ward.[32]

Hewitt's fourth and final stuffed great auk was obtained under curious circumstances. He had heard of a Scottish laird with a mounted great auk and egg for sale. After a protracted correspondence with the owner, Hewitt sent his colleague and fellow oologist, Peter Adolph, to get them for him in July 1948. With a car, a driver and a blank cheque provided by Hewitt, Adolph set off from his home in Kent for the Jacobean-style mansion known as Poltalloch House in Argyllshire, some 500 miles distant. On his arrival, Adolph was greeted by the kilt-attired Colonel Malcolm, 18th Laird of Poltalloch. Pointing to the large round case containing the bird and the egg, Malcolm told Adolph that whatever he was going to offer, it would not be enough. However, when Malcolm stated his price – £500 – it was 75 per cent less than what Adolph had anticipated. An absolute bargain.[33]

As mounted specimens go, the Poltalloch Auk is a particularly good one. It had probably been stuffed by the skilled Soho-based taxidermist, Benjamin Leadbeater. The skin had been purchased by the Malcolm family in 1840, possibly soon after the bird's death, although nothing is known of its previous history. Like all of those who stuffed or illustrated great auks, Leadbeater's aim was to create a lifelike mount of a bird he had never seen. Taxidermists faced exactly the same problem with other extinct birds, most notably the dodo. A combination of misinformation and artistic licence resulted in a public image of the dodo that was too fat for its own good.[34] The great auk suffered similarly, with many specimens overstuffed and obese, while others are overly long and appear emaciated. Whether taxidermists and artists were successful in recreating an authentic auk depended partly on their skill, and in the great auk's case, partly on personal experience with other auks, especially the razorbill. The Poltalloch Auk, along with Hewitt's three other stuffed auks, were among the very best of the 78 surviving specimens of the species.[35]

The great auk enthusiast Irene Kaltenborn pointed out to me that there's another aspect to this. It is that all stuffed specimens and almost all painted images of great auks are birds standing on land, as though to emphasise their desperate vulnerability, a vulnerability that feels almost like a death-wish. There are, fortunately, a few exceptions, including Edward Lear's bird on the water, painted for John Gould's *The Birds of Europe* in the 1830s. Lear later used an almost identical posture as a caricature of himself, as an 'odd bird'.

Edward Lear's cartoon of himself as an ornithological oddity, from a letter to Lord Carlingford in September 1863. (Courtesy of Derek Johns)

Other images of the great auk are distinctly un–auk–like and lack the appropriate 'jizz' (the difficult-to-define 'feel' of a bird). Lear's ability to capture the true spirit, the gestalt or gist of this bird and most other species that he illustrated is a measure of his extraordinary talent.[36] The point is made by looking at illustrations of other auks, notably Atlantic puffins, but also common and Brünnich's guillemots, all of which sometimes seem to pose an illustrative challenge for even well-established artists. This in turn tells us something about the subtle and far from purposeless way in which evolution – natural and sexual selection – has shaped, literally, the bodies of these birds for their terrestrial, aerial and aquatic lives.

Hewitt's concern for the fate of his collections, first voiced in the mid- to late 1930s, was partly a response to the likelihood of war, but also because the tide of disapproval against egg-collecting, which started in the early 1900s, was gathering pace. A certain J. C. Squire writing in 1928 noted that 'even with regard to the common birds there is a perceptible change in the public attitude ... the young egg collector is gradually trained to take one egg rather than the whole clutch ... a growing respect for bird life is perceptible.' The increasingly widespread condemnation of all things oological undoubtedly unsettled Hewitt, fuelling concern that his collections would not be valued.[37]

In his own defence, Hewitt was quick to point out that he was no longer an egg-collector in the same way that he had been and others still were. From the 1930s, armed with his fortune, Hewitt collected other people's collections of eggs already taken. He could not be held responsible for the loss of life so incurred. Indeed, Hewitt was committed to both the conservation of birds at his beloved Cemlyn lagoon and the preservation of the collections he had amassed. Whether Hewitt really understood the *scientific* value of his collections, or whether his concern for their future was merely emotional, is impossible to know.

Like Hewitt, Harold Gowland was also anxious about the growing opposition to egg-collecting. Fearing for his livelihood, in May 1937 he arranged for a solicitor to go through the various statutes passed between 1880 and 1934 relating to the protection of wild birds. He was especially concerned about the rights of the police to stop and search suspects. He then wrote to Hewitt, suggesting that he sit back, light up a cigar and brace himself for a long letter:

> *You have probably heard about that chap Tucker who lives somewhere near Kings Lynn who is 100% anti-collector, and who has started a kind of society to put an end to all the ravages of the collectors.*[38]

To see what Tucker was up to, Gowland got one of his office staff to join Tucker's society. It proved to be an informers' club, with all members encouraged to send Tucker a list of all known collectors in their area. 'In other words,' Gowland told Hewitt, 'there were to be hundreds or thousands of unpaid workers keeping track of the movement of local collectors.' He continued: 'one of the pamphlets issued stated that this new society knew the names and addresses of over 150 large collectors. You are sure to be on the list.'[39]

'The first blow has been struck,' Gowland told Hewitt – and this was the main reason for his letter: 'two chaps from London had been stopped and searched by the police in Cornwall, and found to have two clutches of raven eggs.'

'They phoned Jourdain about the matter, asking if I could help in a certain way which I need not describe here, and asked Jourdain also to write to me on the matter, which he has done. I cannot put in writing the suggestion made except to say that I could not consider it for one moment.' Gowland's statement is ambiguous, and it is not clear whether 'they' refers to the 'two chaps' or to the police. But his point to Hewitt is that 'this stopping and searching of cars is likely to happen to all of us'.

The other point of Gowland's letter, however, was to persuade Hewitt to pay (£5 5s) for Counsel's opinion on the veracity of what his solicitor said, that 'we can be stopped and searched on the highway like a lot of criminals, or whether we can rightly tell the policemen to go to blazes'.[40]

Gowland's concerns came not a moment too soon, for on 2 June 1937, he and his assistant Frank Whatmough were apprehended by the police after taking guillemot and razorbill eggs from Carreg y Llam on the Llŷn Peninsula. 'You know the place, of course, as you were there once with us,' he told Hewitt. 'We had about 200 eggs, all picked specimens including some very decent razorbills.' Gowland argued that he was unaware that the birds were protected. The policemen thought otherwise, and they all went to the police station to check. 'The bobby didn't know one egg from another and persisted in calling the razorbills "razorblades"'. Gowland and Whatmough were given lunch (lamb and mint sauce) at the police station. When one of the policemen took four of the eggs for his sons, Gowland showed him how to blow them. 'The whole idea of being prosecuted for taking guillemots is ridiculous … especially since according to the press there is a scheme to market guillemot eggs for food purposes from Bempton to America … still, if they do fine us the maximum, it will be a bit of a devil.' As it was, no further action was taken. Gowland had escaped – for now.

Later that year, in September 1937, Gowland sent Hewitt another long letter of five closely spaced typed pages on a topic that must have been preoccupying him for years: 'My proposal is that for £250 a year (about £14,000/$18,000 today) I would place myself entirely at your disposal for 6 months of the year ... I would guarantee that you got value for your money ... I know you think quite a lot of me ... why not give it a trial for 3 or ... 4 years ... Just think it over. ... All the best, Captain, and I hope you can reply favourably.'[41]

It was simultaneously sensible and staggeringly naive. Gowland worked like a dog, both for his employer at the shipping company and also at his own egg business. He needed a break, and Hewitt with his limitless funds seemed to offer an opportunity. Hewitt, moreover, desperately needed assistance in organising his collections. On the other hand, Gowland failed to appreciate just how private a person Hewitt was, and the response, of course, was a firm 'no'.

Down but not out, Gowland continued his relationship with Hewitt much as before. They were like lovers, sometimes looking out for each other, sometimes bickering: 'I have not heard from you for ages,' says Gowland – repeatedly. It was a 'master and servant' relationship bolstered by occasional affection and concern. In 1940, Gowland's 43-year-old wife, Emily, who had been ill for some time, died. Within six months, though, he was remarried: 'Olive is much younger than I am, but she's a very nice kid,' and she readily became a substitute mum for Gowland's two children. Curiously, news of Gowland's new wife encouraged Hewitt to reveal that in the past he had had an affair with another Olive. Surprised by this revelation, Gowland responded by saying: 'this is a new name to be associated with you, and as more and more information keeps slipping out, I am afraid you must have been a bit of a bad lad in the years gone by.'

In 1942, Mrs Parry's son Vivian died when the Wellington bomber in which he was a rear gunner was forced to make a crash landing in Yorkshire on its return from a mission over Bremen. On receiving the news, Gowland sent Hewitt a long heartfelt letter of condolence: 'You have seen him grow up from a baby, through all the various stages until he became a man, and during this time you have helped him and guided him just as much as if he had been your own ... Somehow, I have always thought that Viv was your favourite amongst the boys, possibly because he did seem to take more interest in your hobby than Jack ever did.' Among the photographs found after Hewitt's death, there is a poignant image of Mrs Parry kneeling at the side of her son's flower-strewn grave in the tiny cemetery of St Rhwydrus church, a little over a hundred metres from Bryn Aber.

Five years of war seemed to drain Hewitt's enthusiasm for birds' eggs. On 4 January 1946, he sent Gowland a rather bleak letter, saying how disappointed he was with Jourdain's collection, which he had bought on the Reverend's death in 1940:

> Much of the Jourdain collection was no good as such a lot of the data had been destroyed and also someone had surifed [illegible] some of the sets. Many of the books were missing too and all the photographs had been taken − I'm sorry I paid such a price for it as it wasn't worth it.[42]

He continues:

> *I have added very little to my collection for the past six years ...*
> *there does not seem much object in going on with it, and I have*
> *had a very good offer from America ... If I decide to accept this*
> *offer, I should of course give up egg collecting.*

Hewitt adds that he intends to visit California later in the year.
'I may decide to domicile myself there ... as I have lived in
America before and like the country and the people.'

After asking Gowland how his business is going, he finishes
by saying: 'I often think of the times when I used to keep you
up half the night going through eggs – Happy days, I don't
expect we shall ever see them again.'[43]

A week later Gowland replied in a long letter, either
genuinely sympathetic to Hewitt's plight or worried that what
had once been a major source of income was drying up:

> *Reading between the lines of your letter, it looks as if you admit*
> *yourself that your collection is getting too large for you, and that*
> *on account of this, you are losing a certain amount of interest.*
> *This must never happen.*[44]

The winter of 1946–47 was a particularly harsh one in Britain.
For weeks the snow lay deep around windswept Cemlyn Bay.
The cold was intense, especially in a house with no heating,
electricity or running water. Life must have been tough, so it
is not surprising that soon after that dreadful winter Hewitt's
thoughts turned to warmer climes. Initially, accompanied by a
male colleague, Hewitt tried Bermuda, but he didn't like it.
From there he visited the Bahamas and loved it. He and Mrs
Parry subsequently enjoyed a succession of holidays there,
revelling in the warm sun and clear seas. Then, in 1953, he
decided that he would move permanently to the Bahamas.

Now aged 65 (with Mrs Parry now 70), his initial motivation was comfort and health, for he suffered badly from bronchitis each winter. However, on discovering that the 'Bahamas impose no income taxes, capital gains taxes, personal property taxes, or probate of death duties' he was delighted. For someone so utterly averse to taxation and who, between 1947 and 1950, felt he had been 'robbed' by a 'socialist chancellor', moving to the Bahamas seemed a sensible thing to do. In November 1953 they left Bryn Aber for Mount Vernon, a beautiful colonial-style house a few miles from Nassau, set in 67 acres of land. Hewitt paid £33,000 (equivalent to £766,000/$975,000 today) for this property.[45]

Hewitt's migration to the Bahamas marked the end of his collecting mania. He left the Anglesey house and his vast jumble of collections in Jack Parry's capable and devoted hands. He had all his really valuable items shipped out to Mount Vernon, including his four great auk skins and 13 eggs. Mrs Parry's daughter Girlie and son Ken both spent a lot of time with their mother and the Captain at Mount Vernon. This wasn't always easy for Girlie, since it meant leaving her husband, Tom Davies, and their son, Vivian, who was at school, for long periods in Anglesey. As Hewitt and Mrs Parry aged, Girlie and Tom spent more time with them, eventually moving permanently to the Bahamas to help care for them.[46]

Hewitt's time in the Bahamas was exceptionally happy. He loved fishing, swimming and boating, and enjoyed being able to walk from the house onto his private beach. Mrs Parry loved it, too. She was supported by a houseful of servants and two of her children and, during the school holidays, her grandson. Free at last from his compulsive collecting, Hewitt's life was now much more normal than the one he had led at

Bryn Aber. Despite his age, the Bahamas saw him 'recapture the zest of living which had somehow eluded him in middle age'. As his biographer says, 'he became the carefree, vivacious, devil-may-care Vivian Hewitt, reminiscent of his pioneer aviation days'.[47]

The timing of their move into the sun could hardly have been better, since, in 1954, Britain's new Protection of Birds Act made it illegal to take and sell the eggs of wild birds. Vigorously supported by the RSPB, the Act played a key role in boosting their membership. But it was not without controversy for, as well as being directed at adult egg-collectors, the Act also sought to bring to heel the 'swathe of children and young people for whom bird's-nesting was a well-established and routine pursuit'.[48] Many, though, considered this criminalisation of young people ridiculous. Others voiced the benefits of birds'-nesting. The writer, critic and naturalist Geoffrey Grigson chipped in with an article in *Country Life* magazine, pointing out how childhood egg-collecting fostered an adult respect for nature.[49] One 44-year-old woman recorded that her interest in birds'-nesting stemmed from 'an affection for creatures, first of all; excitement of finding pride in knowing something by myself; the indescribable feeling of pleasure at seeing the nest, so intricate, so unbelievable'. The MP Major Tufton Beamish, a member of the RSPB's council, was concerned with the absurdity of a law that meant that every birds'-nesting child was liable to a penalty. Professional dealers in birds' eggs – like Harold Gowland – on the other hand, were a serious problem, he said.[50]

Gowland had resigned from his position at the shipping company a few years previously to concentrate on what he called 'natural history'. To this end, he had promoted his young and eager assistant Frank Whatmough to the position of business partner. Gowland fought the new law and thought he could dodge the issue by focusing on the eggs of *foreign* birds.

But he was naive, for the law recognised only species, not geographic subspecies. Consequently, in 1955, when Gowland was asked by a female customer for the eggs of 'Continental goldfinches', he failed to realise it was a sting. In the eyes of the law a goldfinch was a goldfinch, regardless of whether it was Continental or British. Gowland must have been in the RSPB's sights for decades, and his was their first conviction under the new Act. His house was searched, eggs were taken away, and he was 'found guilty on two summonses of offering eggs unlawfully for sale, and one on summons for actual sale'. He was fined £55 with £50 costs (equivalent to £2,300/$3,000 today). Gowland himself had become a trophy for the protectionists. That was in July 1955; in December the same year, he was convicted again, this time for selling four greenshank eggs. As if that wasn't enough of a blow, Gowland discovered that his young wife Olive was having an affair with his business partner, Frank. Little wonder, then, that at the age of just 57 in 1957, Gowland suffered a fatal heart attack.[51]

Hewitt and Gowland's 30-year exchange of letters ended in June 1955 with news of Gowland's conviction. Fearing Gowland's earlier pronouncement that he too would 'be on the list', Hewitt stepped away. Whereas the sight and touch of eggs and stuffed birds had once breathed new life into him, it seemed that now the very act of leaving all of that behind had exactly the same effect. For more than a decade, Hewitt and his family revelled in the West Indian sunshine.[52]

Until …

Chapter 9

Post-mortem Fate

In February 1965, Hewitt wrote from his Nassau home to Mrs Parry's son, Jack, at Bryn Aber: 'My tummy is still wrong and I am unable to eat anything solid.'

He had been ill for some time but, as always, was reluctant to seek medical advice, or, when he did, refused to accept it. Things eventually became so bad that Hewitt flew to Miami to consult a specialist. The diagnosis was clear but because his condition was incurable, neither he nor Mrs Parry were told. Now cared for by Mrs Parry's widowed son Ken, who lived with them in the Bahamas, Hewitt refused once again to follow his medical consultant's advice and continued to decline.[1]

At Bryn Aber, Jack was concerned about both 'Captain' and his mother, but he was also worried about the health of one of the eight precious parrots that Hewitt had left in his care. The bird – an African Grey that Hewitt had bought 30 years previously – was sick. It died soon afterwards.

In March, perhaps sensing unconsciously at least that the end was approaching, Hewitt arranged for his solicitor to transfer the ownership of Bryn Aber, its contents and grounds to Jack. A major concern as always was avoiding tax and death duties. Hewitt asked Jack to arrange for the estate – the land outside the walls of the house, including the lagoon – to be given to the North Wales Naturalists' Trust. 'The whole thing is to avoid anything in the way of "Real Estate" … It would be a good thing if you would go into this matter immediately on my behalf with the Trust and tell them that I would feel disposed to hand over the Estate … You might ask them outright if this hand-over absolutely avoids death duties on my behalf. I want absolute assurance on this point.'

The person Jack dealt with at the Trust was a local ornithologist, Pat Venables, who later played a crucial role in determining the fate of Hewitt's collections. By late April, Venables had persuaded the Trust to keep the estate as a Wild Bird Sanctuary. And there was more good news – for the first time, a pair of grey herons bred within Bryn Aber's walls. On being told, Hewitt wrote back to Jack saying: 'I am so very delighted … I had always hoped against hope that this might happen, but I never expected it.'

Hewitt's health continued to deteriorate through April and May, so he decided to return to Britain, telling Jack to expect them at London Airport (shortly to be renamed Heathrow) on 18 June. It was Ken who had to get Captain and Mrs Parry onto the plane. Both were unwell. Writing to Jack, Ken said: 'Mother is quite lost and lives within a world of her own making, and I'm sure doesn't realize just how ill Captain is.'[2]

As Hewitt's doctor later commented:

How often have I observed this … yearning for home when death is approaching … Seeing him shortly [after the journey] I would not have thought him capable of travelling the shortest of distances, let alone the … journey from the Bahamas. He was literally skin and bone … unrecognizable from the Vivian Hewitt I had known – the sunken eyes, the pathetic expression, the hopelessness and despair associated with physical weakness.[3]

Unaware of quite how close he was to death, Hewitt thought that he would soon be returning to Nassau. Over the several weeks that Hewitt deteriorated once he was back at Bryn Aber, his doctor of over 30 years, William Hywel Jones, reflected on his own helplessness to avert the inevitable. Musing on Hewitt's extraordinary life, he later wrote: 'What I find hard to understand is the complete transformation in his aspirations after the acquisition of his wealth … to a life of relative isolation and limited desire … so much started, so little completed.' Stubborn, selfish and set in his ways, Hewitt maintained his relationship with Mrs Parry and the four children, acting, as Hywel Jones contentedly observed, 'as one family', which may have been literally true.[4]

Oesophageal cancer is among the deadliest of all cancers, and it was obvious to Hywel Jones that Hewitt should be in hospital. But as intransigent as ever, he refused. Luckily, a Liverpool surgeon whom Hewitt knew was staying in his holiday home nearby. Hywel Jones persuaded him to come to Bryn Aber and, shocked by Hewitt's state, the surgeon made arrangements for his immediate transfer to Liverpool Royal Infirmary. But only after four more days of prevarication was an ambulance able to take him there, with a police escort, because his condition was now critical. Surgery finally allowed Hewitt to drink and gave him a further four weeks of life. Discharging himself from hospital, his final days were spent with Girlie and Tom in their comfortable home at Tŷ'n Llan, close to Bryn Aber, where Hewitt died on the morning of 18 July 1965.

In leaving his collection of birds' eggs and skins to Jack, Hewitt undoubtedly felt they had financial value. Jack thought so too, initially. However, on being informed that they could not legally be sold (which was actually untrue, but was the prevailing opinion), Jack's accountants suggested he 'tip the whole lot over the cliff'.[5]

Getting wind of this, Pat Venables was concerned. A week after Hewitt's death, he wrote to David Wilson, secretary of the British Trust for Ornithology (BTO), whose headquarters were in Tring, Hertfordshire. Venables hoped that Wilson could persuade the BTO to accept the collection. The initial focus of Venables's and Wilson's efforts was to secure Francis Jourdain's notebooks – which they knew Hewitt had – for the BTO's library.[6]

Recognising that swift action was needed, Wilson drove from Tring to Anglesey on 20 September to meet Jack Parry. The scale, complexity and urgency of the situation surrounding Hewitt's collections was immediately clear. The quantities involved were enormous, filling not only the house from floor to ceiling but more than 20 wooden sheds within Bryn Aber's walled perimeter. To avoid death duties, Jack was desperate to be rid of his ornithological inheritance, and as soon as possible.

In a solemn ceremony involving Wilson, Venables and a Hewitt family friend, conducted on the doorstep of Bryn Aber, Jack Parry handed over ownership and responsibility of the collection to David Wilson. Venables and Wilson were not allowed in the house, because Jack's mother, Hewitt's lifelong companion Nellie, 'had gone upstairs' – that is, she was no longer *compos mentis* – and was 'prone to shouting at strange men'.

Persuaded by Wilson, the BTO agreed to take Hewitt's collection to 'save it for science' – so they said. In truth, the director saw it more as a way of 'turning the material into cash' to enable them to build a much-needed extension to their headquarters. Wilson wasn't happy about this. Nevertheless, four days after his visit he wrote to Jack Parry:

This is to confirm the agreement made last Monday [20 September 1965] *between yourself and Mr Venables and myself, and I confirm that the BTO will accept the various Jourdain notebooks and documents. I also confirm that the BTO will accept the egg and skin collections as we agreed. We shall arrange for collection at the very earliest opportunity when all formalities have been completed.*

Knowing that Jack Parry was desperate to complete the transfer with as little fuss as possible, Wilson instructed the BTO's solicitors to contact Parry saying that this was a 'private gift, no publicity; BTO needs proof of ownership and no liability for death duties; please pass this to your solicitor and ask him to confirm ownership and no outstanding liabilities'.[7]

In October 1965, just four months after Hewitt's death, Wilson sent a confidential note to his boss, Dick Homes, president of the BTO: 'There are also some eggs and skins and possibly a fairly good library ... The Great Auk eggs and the like are overseas ... An all or nothing offer, as the present owner wishes to get rid of everything in one go and as soon as possible.' Wilson added that he had checked with Nature Conservancy and, since everything was collected before 1954 (i.e. before the 1954 Protection of Birds Act), no permit was needed. 'I will keep you in touch with developments. The whole thing is quite extraordinary, and even if nothing eventually materializes it has been a most interesting experience dealing with the old boy.'[8]

Wilson was clearly excited. Having met Jack Parry and seen the outside, if not the inside, of Fortress Bryn Aber, he could only imagine the vast treasure trove within.

Despite Parry's request for secrecy, news of Hewitt's death and the imminent disposal of his collection had leaked. Museum curators started to gather like flies around a carcass, hoping for some of the action. Among the first was Reg Wagstaffe of Liverpool Museum, who had asked another Anglesey resident, the bird artist Charles Tunnicliffe, to make enquiries on their behalf. There were others too, including the Smithsonian Institution in the US. In response to a request from Leicester Museum, Wilson replied:

As far as I know the best items (Great Auks, etc) and a considerable part of the rest of the collection are overseas and probably unlikely to be returned.

But it was all taking too long. Venables wrote to Wilson in December 1965, telling him that Parry was frustrated. His marriage was 'on the rocks' and 'his old mother has now gone completely batty and will not have another female near the house. Parry is too tender-hearted to send her to the bug house, so he remains single ... Well, do try to hasten your collecting of the collection'.[9]

In April 1966, James Macdonald of the Natural History Museum wrote to David Snow, director of research at the BTO:

I have very slight knowledge of the Hewitt egg collection ... Among other things there is believed to be about 10 or a dozen Great Auk eggs. We tried some time ago to get confirmation of his specimens in connection with a publication on the known [great auk egg] specimens which we are at present printing, but I think he was past dealing with correspondence ... I have an idea that Hewitt's was probably the largest and most complete private Egg Collection in the world – even greater than Kreuger's at Helsinki. Hewitt was very wealthy and never spared money in getting specimens he wanted.[10]

In May 1966, after several huge trucks full of eggs and cabinets of skins had made their tortuous way from Cemlyn Bay to the basement of the Natural History Museum in Tring, Wilson wrote to Venables:

Only concern is importing woodworm into the museum – but they soaked everything in 'Rentokil'. Nothing as yet in the way of startling new discoveries but vast quantities of data. Many interesting letters not for publication. Apparently collecting eggs whether in the nest or in a cabinet was considered fair game by some collectors [meaning that they often stole from each other?]. *Ground floor* [of Bryn Aber] *has been emptied and much from upstairs. The old lady* [Mrs Parry] *went to the West Indies on 24 April. Jack is a new man.*[11]

Wilson in turn, in a letter dated 20 May 1966, brought his boss Dick Homes up to speed:

Two van loads already at Tring and two more by 4 June. Cost £60/van and c. £500 in total. Specialises in birds of prey, petrels and seabirds in general. Collection contains the Jourdain collection – probably one of the best. Aim of the collection [i.e. for the BTO] *... possibly to replace the Rothschilde* [sic] [*i.e.* Walter Rothschild's collection in the Natural History Museum].

Wilson told Homes that the collection had been offered to the BTO, 'but the wish has been expressed that he [Jack Parry, echoing Vivian Hewitt] does not want it to go to the British Museum and that, as far as possible, the collection remains intact for the purpose it was formed'. The BTO were unable to respect Hewitt's wishes, and it may have been for this reason that David Wilson decided not to pass everything on to them.

Amid the vastness of Hewitt's collections, everyone's eyes were on the most valuable specimens – the great auk skins and eggs. On 22 March 1967 Wilson wrote to Jack Parry to ask:

Do you know what is going to happen to the Great Auks? I know of someone (a museum not a private individual) who would like to have the opportunity of making a bid for one skin and one egg. If they are to be sold in America, will you be in a position to have first refusal?

Parry replied:

> *Now, to the Great Auk(s) ... I have no say in the disposal of the eggs or skins ... my brother Major Ken Parry* [in Nassau] *... is handling all such matters.*

On 26 July Wilson harried Parry again about the great auk eggs:

> *When I was last with you, you said that one of the directors of Spink & Sons* [sic] *has been over there* [to Nassau]. *Do you know if they are bringing the other items back to this country for sale? If so, perhaps, the eggs could come back in the same way as they went out there.*

Other museums were also after the great auk eggs and skins, but Wilson fobbed them off by saying that they were 'taken by Captain Hewitt to his home in the Bahamas and there seems to be little prospect of their return'. The director he mentioned was David Spink, chairman of Spink & Son, a long-established (since 1666) London auction house for selling coins, porcelain, medals and stamps, all of which Vivian Hewitt collected. He had sought counsel from David Spink, and over the decades spent a fortune with the company.

On 5 June 1968 – almost three years after Hewitt's death – the final truckload of material arrived at Tring. Dick Homes had decided that they would sell all the mounted specimens except the birds of prey, and would deal with the eggs later: 'We are not anxious to be commercial about this ... but we have incurred over £500 expenses in rescuing the collection.' He was keen to emphasise that the BTO had been established to promote the study of *living* birds rather than their remains, and Hewitt's collection was both outdated and of little scientific value.

Once Hewitt's material was at the BTO's headquarters, David Wilson and another staff member, Kenneth Williamson, spent weeks, often late into the evening, attempting to document the seemingly endless array of cabinets and cases. In total there were some 15,000 skins and around half a million eggs. Their difficulties were exacerbated by the fact that Hewitt himself had never catalogued his collection.

One way the BTO considered dealing with Hewitt's collection was to 'retain a major collection of the eggs of British birds'. In response to this suggestion, the Natural History Museum's Curator of Birds, Colin Harrison, pointed out that not only would this require 300–400 new egg cabinets, it would also take 20–30 years to put the collection 'into its proper shape for research purposes'. He added 'I think that the Trust [BTO] should give serious consideration to the question of whether the initial work and expense required in sorting, identifying, and arranging a large quantity of material will justify the ultimate acquisition of a collection of British eggs which may not contribute much to the work of the Trust.' An embarrassment of riches.[12]

In December 1968, Wilson drafted a round-robin letter advertising the disposal of Hewitt's bird skins: 'Offers should be sent to me in writing by 31 January 1969.' Several UK museums came to the feast. They removed material, including some of Hewitt's eggs that the British Museum was not interested in. The biggest coup for the BTO, however, was the appearance of an American millionaire, John Eleuthère du Pont, who was fortuitously in the process of establishing his own natural history museum in Delaware.

Du Pont, like Hewitt, had inherited a fortune – in his case from what was originally the family's gunpowder business, but later from its chemicals and cars. The du Pont family was among the wealthiest in the United States, and John du Pont was a respected ornithologist who was later (in 1973) awarded a PhD from Villanova University, Pennsylvania. As well as eggs, du Pont also collected stamps and mollusc shells. Outside his obsession with collecting, he was fanatical about athletics and wrestling. After buying virtually all of Hewitt's collection once everyone else had taken what they wanted, du Pont's museum eventually opened in 1972. The £10,000 that du Pont paid represented a huge contribution to the proposed extension to the BTO's headquarters. Sadly, and remarkably, it seems that no records were kept of what du Pont took, but when I visited his Delaware Museum of Nature and Science with my colleague Bob Montgomerie in 2016, I was amazed to see literally thousands of guillemot eggs. Almost all of them were originally from Bempton, and were from the various collections Hewitt had purchased. And as I rummaged gently through the trays of eggs, I saw among them the familiar, scrappy pencil notes in Hewitt's hand.

Initially, the BTO felt that du Pont should get *all* of Hewitt's eggs but, when word of this got around, British oologists made a strong case that Jourdain's collection, which Hewitt had bought in 1940, should remain in the UK – which it did.[13]

John du Pont was an unpleasant man. When he visited Tring to buy the Hewitt material, those present on the day felt that he and his bodyguard–cum–chauffeur were a threatening presence. Years later, in 1996, du Pont shot and killed Dave Schultz, his wrestling coach, and spent the rest of his life in prison, dying in 2010.

I can only imagine David Wilson's dismay as he watched Hewitt's huge hoard of eggs and skins ransacked by all and sundry. As was discovered only after his death, Wilson kept all of Hewitt's papers and some of Hewitt's favourite specimens in his home. I was to inherit some of these for our university Zoological Museum in Sheffield.[14]

There was a lot of scurrying around after Hewitt's death – especially by David Wilson – to locate Hewitt's hugely valuable great auk eggs and skins. Wilson had been told previously by Jack Parry that David Spink had been out to Nassau. Putting two and two together, Wilson thought it likely that Spink had brought the great auk specimens back to London. Throughout his time as a collector, Hewitt had purchased a vast amount through Spink, and although best known as a dealer in antiques and art, Spink & Son was the family's natural and trusted choice for disposing of Hewitt's great auk relics and other ornithological material. Wilson's suspicions were confirmed when, by chance, he came across an article in the *Illustrated London News*[15] announcing Spink & Son's forthcoming sale of Hewitt's great auk specimens. Wilson travelled to London on 9 June 1968 prior to the sale to meet David Spink. He was shown the eggs but not the skins, which he said were elsewhere in the building. Spink told him that the great auk eggs were to be retained by the family. This was a lie.

Financial constraints resulted in David Wilson being 'let go' by the BTO in 1970, at the age of 44. He swiftly metamorphosed into one of Britain's best-known book dealers, specialising in books on birds and British islands (including St Kilda). He transformed the upstairs of his modest Aylesbury home, including the roof space, into a storage facility, where, among much else, he retained the 'special' items from Hewitt's collection. Wilson died in 2020, aged 93.[16]

The cycle of acquisition, storage, death and disposal went on and on. Hewitt amassed his collections from earlier collectors and, once acquired, hoarded them with barely a second glance at what he had collected. Then, as we have seen, they were rather carelessly disposed of after his death – except for his favourite items, retained by David Wilson. Having seen these for myself in Wilson's home, I was impressed by just how astute Wilson had been. I was also relieved to learn that Wilson wanted to secure the memory of Vivian Hewitt. He kept all of Hewitt's papers, letters and photographs, items that could so easily have been dumped. Fortunately, Wilson's heirs were keen to pass on items of scientific and historic interest to appropriate institutions, both from Hewitt's collection and also from other material Wilson himself had accumulated.

One of Hewitt's items retained by Wilson was a replica great auk made by the taxidermy company, Rowland Ward. Hewitt had purchased this replica auk from Ward in 1922, at the time that he was becoming interested in ornithology. He paid £8 10s for it, plus 12/6d for packing (equivalent to £420/$560 today). From about 1917, one of Ward's employees, George Griffin, assisted by Alfred Bannister, created fake great auks. They did this better than most other manufacturers, by attaching body feathers from other auks, one by one, onto a papier-mâché carcass covered in a flexible layer made from a rubber solution – 12,000 feathers in all. The wings were either from a guillemot or a puffin, trimmed to match those of a great auk, and the head and beak were cast from plaster and then painted. Other companies that made great auk replicas used patches of skin from smaller auks sewn together and rather conspicuously stitched onto a false body.[17] At a sale of Wilson's belongings in 2021, Hewitt's fake auk was expected to sell for just a few hundred pounds, but instead went to a French collector for a staggering £25,000.[18]

Once immensely popular, from the 1960s most taxidermy was neither desirable nor politically appropriate. It was for this reason that Hewitt's vast collection of stuffed birds was effectively rejected by the British Trust for Ornithology. Taxidermy specimens continue to cause unease and offence; as I was writing this chapter, I heard of a school that, on discovering it owned such a collection, simply burnt the lot. One hopes that their collection did not include a great auk.

What of Hewitt's genuine auks, and his 13 eggs – what became of them?

Chapter 10

Witch Hunters

While preparing his biography of Vivian Hewitt in the 1970s, William Hywel Jones wrote to the British Trust for Ornithology, asking for a list of all the skins and eggs they had inherited from Hewitt. He was hoping to include the information in his book. The Trust demurred, saying: 'Regrettably, it would be very difficult for us to provide you with the information you require.' When Hywel Jones later visited their headquarters in Tring and saw for himself the scale of the problem, he understood the impossibility of providing a list of specimens.

At the end of his visit, and as he prepared to leave, Hywel Jones asked the Trust's director, Jim Flegg, what standing Hewitt had held as an ornithologist. I can almost hear Flegg's keen intake of breath as he struggled to muster a diplomatic reply. Prefacing his response with the fact that he had never known Hewitt, Flegg ventured that the best posthumous assessment would need to be based on his collection, adding, 'but … as an egg-collector the Captain must be ranked very highly indeed'.[1]

Damned by faint praise. The truth is that, as an ornithologist, Hewitt had almost *no* standing. In Flegg's eyes, egg-collectors, and especially high-ranking ones like Hewitt, were the devil incarnate – and their activities were the antithesis of ornithology.

In accepting Hewitt's collections, the BTO found itself floundering between the devil and the deep blue sea. On the one hand, egg-collecting was now illegal, and 'real' ornithologists vilified egg-collectors for the damage they inflicted on individual birds and their populations. On the other, the BTO seems to have had a vague awareness of the possible scientific value of the eggs. Hewitt's stuffed birds, however, were different. Such taxidermic trophies had simply gone out of fashion and, in their large glass cases, were inconveniently bulky. Above all, stuffed specimens were felt to have no scientific value. In a practical sense, and despite David Wilson's monumental efforts, the BTO was simply overwhelmed with the enormity of its inheritance.

Assessing anyone's legacy, or afterlife, can be difficult. The criteria we use depend on many factors, including ideology, economics and our personal viewpoints. Hewitt's legacy is especially challenging because of the 1954 Act that made the taking of wild birds' eggs illegal. As one of my ornithological colleagues said to me, many of the men who collected eggs – and they included doctors, lawyers and businessmen – went to bed one evening as respectable members of society and woke up the next morning as criminals. This did not apply to Vivian Hewitt because the eggs in his collection were all obtained long before 1954, but to many people even those who *owned* eggs were tarred with same brush, shaping the way Hewitt and his collections were viewed after his death.

William Hywel Jones probably knew Hewitt better than anyone else outside the family, but he knew very little about birds or the value of scientific collections. It is unfortunate, then, that it is he – through his biography – who, so far at least, has moulded our memories of this 'modest millionaire'. Hewitt merits a more nuanced and scientifically informed legacy. He was certainly an avid collector during the 1920s and 1930s. After inheriting his fortune he metamorphosed into a collector of collections, and ceased to pose any threat personally to bird populations. Indeed, during the 1930s he became an ardent bird protectionist, and saw himself increasingly as the custodian of other people's collections.

In an albeit unfocused kind of way, Hewitt recognised that his collections had both historic and scientific value.

His antipathy towards museums was unfortunate, for I suspect that had they been the recipients of his collections, the contents might have been better preserved and valued. As it was, the dismemberment and dispersal of Hewitt's collections was, in hindsight, shameful. Hindsight, of course, is a wonderful thing, and I have tried to put this into an appropriate context: there was simply too much material, too little time, too few resources and too much ideology. But there was also too little vision. To learn from history means learning from our mistakes. I'd like to think that such an event would not happen today, but I realise – again with the benefit of hindsight – that it could do so all too easily. I suppose, in a way, that I am simply saddened by the loss of data and the fact that all those eggs died in vain.[2]

Hewitt's most highly prized possessions – his great auk eggs and skins – escaped the careless museum frenzy and instead were put up for public auction.

The catalogue Spink & Son produced announcing the sale of Hewitt's four great auks and 13 eggs was written by one Richard Ford, about whom little seems to be known. Each item was illustrated, and as well as auk material, the sale included the eggs of two other extinct birds once owned by Hewitt; an elephant bird and a 'Russian' ostrich. It seems somewhat naive to have offered the great auk specimens all at once, rather than selling them off gradually. By advertising them together, Spink essentially flooded the market.[3]

As far as the stuffed auks were concerned it didn't seem to make much difference, as all four sold. The first to go was the Los Angeles Auk, purchased by the Los Angeles County Museum of Natural History for $12,600 in 1970. Next was the Clungunford Auk, bought by Britain's Birmingham Museum and Art Gallery in 1971 for £9,000. That same year, the National Museum of Wales bought the Poltalloch Auk that Peter Adolph had secured at a knock-down price for Hewitt in 1948, together with one of the eggs, known as Vivian Hewitt's Egg. Together with Hewitt's elephant bird egg, the sale price was £11,000. The fourth and final bird, the Cincinnati Auk, was purchased in 1974 by the Cincinnati Museum of Natural History, together with the Cincinnati Egg, for $25,000.[4]

Of the remaining 11 great auk eggs, three were sold to three unknown purchasers, for unknown sums at unspecified dates. Malcolm's Egg (the cracked one from Poltalloch House) and Captain Cook's Egg were both sold sometime between 1970 and 1976, and Wallace Hewett's Egg was sold sometime between the late 1970s and early 1990s.[5] It seems incredible, but Spink & Son apparently kept no records, so the whereabouts of these eggs remains unknown.[6]

What of the eight eggs that remained unsold? After he left the BTO in 1970, David Wilson continued to attend the Trust's annual conference in his role as a book dealer. As well as buying and selling books, he built up his own collection of rare and beautiful bird books. Like Vivian Hewitt, he had a passion for islands, with St Kilda being Wilson's favourite. He was a long-standing member of the St Kilda Club, whose aim was to protect all aspects of this beautiful island, one with a long human history that was inextricably intertwined with its seabirds – including the great auk.[7]

Through his book business, Wilson got to know many other bibliophiles. One of these was a Scottish doctor, John (Jack) Gibson of Foremount House, Kilbarchan,

Renfrewshire. Charming, suave and popular with his patients, Gibson had a passion for natural history and islands, and especially in his case the island of Ailsa Craig off Scotland's west coast. In the early 1970s, when in his late forties, Gibson started to create what he called the Scottish Natural History Library at his home. His ambition was to build a collection of everything that had ever been published on the natural history of Scotland. This visionary idea attracted official support, and Gibson secured several substantial grants. The funds he obtained from the Scottish government and foundations such as the Carnegie Trust enabled him to purchase a number of particularly valuable volumes. When the Royal Society of Edinburgh was forced to dispose of its library, Gibson persuaded them to transfer all of their natural history books to his. Delighted by his success, Gibson wrote in 1984: 'With holdings of some 40,000 volumes [the Scottish Natural History Library] is now undoubtedly the largest separate collection of Scottish natural history books and journals.'[8] Another vast collection, and another eccentric.

Gibson was happy to announce his ability to amass books, but was reticent about his passion for birds' eggs. Like Hewitt and many others, Gibson purchased eggs from Harold Gowland, who, on 29 November 1954, wrote to Gibson to say: 'In view of the new law which comes into effect on Wednesday [1 December 1954], I think it would be unwise to send anything by post which is affected by the law as maybe the witch-hunters will be keeping check on dispatches ...'[9] Through his fascination with eggs, Gibson became an honorary curator at the nearby Paisley Museum. In 1958, Dr Ian Pennie, a keen birdwatcher, was given an important collection of eggs – including clutches of peregrines, ravens and greenshanks – once owned by Alexander Macomb Chance (brother of the cuckoo enthusiast, Edgar Chance). Pennie let it be known that he wished to find a home for the eggs and was duly contacted by Arthur Clarke, assistant keeper of birds at the Royal Scottish Museum, who offered to take them. Pennie wrote back to Clarke, saying that the only reason he had not already offered him the eggs was that he had been told by 'someone familiar with the museum – I cannot remember who it was – that you had all the British eggs you required', but that 'I have heard indirectly that Paisley Museum is also interested'. Clarke replied, saying that they were happy to help smaller museums, like Paisley, which 'in particular has had much from us owing to the enthusiasm of a Dr Gibson who generously devotes much of his spare time and money to that institution. I will certainly let him know about this collection.' Gibson acquired the eggs, and forgot – it seems – to pass them on to the museum, for they were found in his house after his death.[10]

Despite its potential as a nationally important resource, there was a major problem with Gibson's library in that only *he* had access to it. Those who wrote requesting a research visit typically received a polite reply explaining that the library was undergoing renovations (or something similar), and was inaccessible. The number of people that succeeded in gaining access during Gibson's lifetime could be counted on the fingers of one hand.[11]

It was through David Wilson in the early 1970s that Jack Gibson got to hear about the eight unsold great auk eggs resting in David Spink's safe. Gibson immediately set off for London, dreaming that he might be able to buy them. His hopes were dashed on being told that the eggs were now the property of the wife of one of the company's directors – actually Spink's wife, Dorothy – and that she was living in a commune in France, and was incommunicado. Spink nevertheless showed Gibson the eggs that he had loose in a drawer. Great auk eggshells are pretty robust, but Gibson was worried that, rolling back and forth each time Spink opened the drawer, they might be damaged. He persuaded Spink to allow him to wrap each of them in bubble wrap. Disappointed and frustrated that he was unable to purchase these eight ultimate objects of oological desire, Gibson made a point of visiting Spink each year to check on the eggs and to reaffirm his wish to buy them. Remarkably, after almost 20 years his persistence paid off, and in 1992 Gibson was finally told by Spink that he could buy all eight eggs. The sum was a remarkable £14,000 – just £1,750 each. A truly astonishing bargain. Presumably Spink simply wanted to be rid of them.[12]

In September 1992, motivated by the boom in molecular studies and the value of museum collections as a source of DNA, the British Ornithologists' Union ran a conference at the University of Liverpool on the history of ornithology. Stupidly I did not attend, mainly, I think, because my interest in ornithological history was relatively undeveloped then, and took second place to my much more mainstream biological research on avian promiscuity.[13]

One of those who did attend was Bob McGowan, curator of birds at the National Museum of Scotland in Edinburgh. Bob later told me that during the meeting he was approached by someone he did not know. The unknown man told McGowan somewhat conspiratorially, in a heavy Scottish accent, that he was about to acquire eight great auk eggs, which he would donate to Bob's museum. This person turned out to be Jack Gibson, and he swore the stunned Bob McGowan to absolute secrecy. It was a proposition equivalent to someone surreptitiously offering Arthur Sackler a set of priceless Chinese treasures – for free.

Thrilled at the prospect of this donation, McGowan told Gibson that such a gift would make his name in ornithology, to which Gibson replied that he had already made his name in ornithology; possibly true, but maybe not quite in the way he thought. That Gibson had a high opinion of himself is evident from a letter he wrote to a wealthy philanthropist in 1986, asking for funds to purchase John Gould's personal set of 47 imperial natural history folios. Gibson described himself in that letter as 'one of the most distinguished Scottish naturalists of all time'.[14]

By early 1993 the great auk eggs had made the long journey from Spink's offices in London to Gibson's house in Renfrewshire. In November that year, David Wilson travelled to Scotland to see them. In the note he later wrote for himself, he said: 'I was the first person to be told of the acquisition and shown the eggs in recognition of my first drawing Jack's attention to them some 19 years ago … My main impression was the unreality of the occasion … it was difficult to comprehend that the eggs were real.' A week after his visit, and still in a state of shock, he wrote to Gibson, saying: 'I still cannot believe I have actually seen and handled the contents of your magic boxes.'[15]

News of Gibson's extraordinary acquisition percolated into parts of the ornithological community. Bill Bourne knew – which is why he fobbed me off when I asked about them around the same time (see p. 13) – and the other person who knew was Errol Fuller, who at the time was writing his great auk *magnum opus*.

Obsessed by the great auk since childhood, Fuller knew more about its history and specimens than anyone else. His aim was to produce a volume that was both beautiful and comprehensive. The publishers that Fuller approached deemed the book to be too expensive to produce, because of the large number of colour illustrations Fuller wished to include. As a result the book was eventually self-published. Fuller's plan was to include images of all extant great auk specimens and eggs. To do so was no mean feat, since they were scattered across the museums of the world or in private ownership. Gibson was the last bastion of oological resistance. When Fuller asked if he could visit him to photograph his eggs, not only was Gibson reluctant, he was furious with David Spink for telling Fuller that he had them. But Fuller was both determined and persuasive. After a telephone conversation, during which Gibson discovered that Fuller was the author of one of his favourite books, *Extinct Birds*, he suddenly acquiesced and agreed to Fuller's request. A residue of reticence remained nonetheless, for when Fuller travelled from his home south of London to Scotland to photograph the eggs in October 1997, Gibson had transferred them to his surgery, rather than allowing Fuller into his home.[16]

Jack Gibson with his eight great auk eggs at his home in Scotland.
(Courtesy of F. Gibson)

Gibson's eight great auk eggs displayed. (Hewitt Papers)

But that meeting helped to forge something of a friendship, albeit fragile, between Fuller and Gibson. This was to prove crucial to our understanding of the subsequent fate of these eggs.

Just as he had done with David Wilson and Bob McGowan, Gibson told Fuller that he was going to give the eggs to the museum in Edinburgh. Accordingly, when Fuller's magnificent book was published in 1999, each of the sections describing Gibson's great auk eggs included a statement that they are 'one of the group of eight eggs he [Gibson] is presenting to the National Museum of Scotland'.[17]

It never happened, and to all intents and purposes the eight eggs simply vanished.

In the early 2000s, during one of their infrequent telephone conversations, Gibson asked Errol Fuller if he knew of anyone who might be interested in buying the eggs. Fuller was taken aback, but understood why Gibson had changed his mind, for they both shared Hewitt's reservations about museums. Fuller also wondered whether this might be an opportunity to buy a great auk egg himself. He was able to suggest a potential purchaser: the extremely wealthy Sheikh Saud Al Thani,

Sheikh Saud Al Thani. (Courtesy Errol Fuller)

Minister for Art and Culture in Qatar.[18] Fuller had previously helped the Sheikh with purchases for the country's incipient natural history museum in Doha. In fact, Fuller had published a book illustrating some of the material that would feature in that museum. With limitless funds, Sheikh Saud was like a combination of Vivian Hewitt, Arthur Sackler and John du Pont on steroids. Exuding a powerful sense of entitlement, he had a discerning eye, an unquenchable acquisitional thirst and bottomless pockets.

He seemed like the perfect person to purchase Gibson's eggs, but it wasn't straightforward. First, in 2010, Sheikh Saud asked Fuller to buy the eggs from Gibson on his behalf. He did so, using £80,000 borrowed from the bank. But before Fuller could be reimbursed, fate intervened. Sheikh Saud fell foul of his cousin the Emir, the country's leader; he lost his position and was thrown into prison. He died, aged just 48, in 2014, reportedly of a heart attack, leaving Errol Fuller with seven great auk eggs and a hefty overdraft.[19]

Only one of Gibson's eight eggs – Vicomte de Bardes Egg No. 1, known also as Lord Garvagh's Egg – made it to Doha, where it is now part of the collection of the Qatari Ministry of Culture.

In an effort to recoup his outlay made on Sheikh Saud's behalf, Fuller subsequently sold five of the seven eggs. Two of these – Alfred Newton's Hunterian Egg and Dawson Rowley's Egg – were sold in 2010 to an unnamed purchaser for an undisclosed sum. Yarrell's Egg was also sold that same year, under similar circumstances. In 2017 Bowman Labrey's Egg – the one my colleague Graham Axon repaired – was sold to a Swedish buyer. Lady Cust's Egg went in 2022 to an acquaintance of Fuller's for £45,000, but within just a few months (in January 2023) it was sold at Sotheby's for around £100,000.[20]

I asked Fuller about the people who bought these eggs. Rather to my surprise, only one was an egg-collector, and, as I was later to discover, had spent time in prison as a result. The others, he said, were individuals that appreciated the eggs for their historical value and the fact that the great auk was extinct.

This leaves Fuller with two eggs. Poignantly, perhaps, these are Jack Gibson's Egg, and Dr Dick's Egg (see p. 156). Driving from Surrey to Errol's home in March 2022, my satnav told me that I had reached my destination. Yet it was unclear which of the close-packed houses was Fuller's. On looking up towards a third-floor window, I saw the familiar shape of an egg cabinet and knew I had arrived. There was a curious resonance in visiting Fuller's house; crammed with museum cabinets and specimens that Fuller had rescued, it reminded me of a tidier version of how I imagined Hewitt's home to be. To some it would be a house of horrors: a stuffed orangutan, numerous gigantic fish heads, vast fossil dinosaurs and endless cases of stuffed birds. For a museum enthusiast like me, however, it was an Aladdin's

Cave – a cabinet of wonders. I was shown and allowed to hold both the great auk eggs.

Fuller's revelation to me about Hewitt's eight missing great auk eggs fills in some important gaps, but, since five of those eggs have subsequently been purchased by unidentified collectors, they are *still* missing. Errol Fuller is the only person who knows who has them, and those eggs remain inaccessible to any researcher wishing to examine them. It is also possible that Fuller may eventually sell his two remaining eggs, so that they too disappear. And, of course, the five Hewitt eggs that David Spink sold in the 1970s without recording any details have also disappeared. Lost eggs are, very occasionally, brought back from the dead. One known as Méezemaker's Egg – *not* one of Hewitt's eggs – was last seen in the 1960s, but it re-emerged in 2021 at Sotheby's sale rooms, where it was on sale for $150,000–$180,000 but failed to attract a buyer.[21]

Of the 13 great auk eggs once owned by Vivian Hewitt, eight are currently in private hands and are – to all intents and purposes – invisible and inaccessible. Only the three in the public museums of Los Angeles, Cincinnati and Cardiff are 'accessible'. The two owned by Fuller may also be accessible, but the rest are even less get-at-able than they were when they were in Gibson's care. From a scientific perspective, the tragedy is that, had Jack Gibson donated his eight eggs to the National Museum of Scotland as he originally claimed he would, they would have been both accessible and safe. On hearing that Gibson had sold the eggs, one ornithologist was heard to respond: 'What a shit!'

Why did Gibson change his mind about donating the great auk eggs to the museum in Edinburgh? For years he had said that his plan was to give the eggs and 'his' library to one or other Scottish public institution. Like all collectors, including Hewitt, Gibson wanted his collection (and especially his library) to remain intact. Such a sentiment is understandable, but it is also unrealistic. Most institutions simply do not have the space or resources to keep duplicates of material they already have, and cannot therefore agree to keep collections intact. Of course, in this case they could have sold *their own* duplicates, not Gibson's, in order to obtain his collections. Did they not think of that? Disappointed and offended that his request could not be granted, Gibson held on to his library and the eggs.

After Gibson's death in 2013, the Scottish Natural History Library, so carefully built up over so many years and often with public funds, was sold privately. Whether the reluctance of certain Scottish institutions to keep Gibson's library intact influenced his decision not to give his great auk eggs to the National Museum of Scotland is unclear. In 2012, Gibson told the then-Chair of the Scottish Ornithologists' Club's Waterson Library, David Clugston, that because of the 2008 global financial crisis, he needed cash to pay for his cancer treatment and to ensure that his wife would be provided for after his death. It was perhaps for these reasons that he decided to sell his eggs rather than donate them to science.[22]

Even though eight of Hewitt's 13 eggs are effectively lost from view, all four of his great auk specimens remain safely ensconced in their respective public museums. The birds have been dead for almost 180 years, but their lives are far from over.

Chapter 11

Afterlife Lessons

I am sitting beneath black cliffs on a tiny and remote island, watching white waves break onto a rocky shoreline. Out of the surf a great auk emerges, horizontal at first, but quickly standing upright as it gains a foothold. After glancing quickly to each side, the bird begins to walk towards the cliffs in an erect and human-like posture. Others soon follow. I watch transfixed as the birds waddle determinedly towards their breeding sites amongst the boulders at the base of the cliff.

My euphoric daydream is demolished by the stomach-churning 'WAAH' of a huge, resentful great black-backed gull swooping viciously at me from the sky.

The island I'm on is one of a group of six known as the Gannet Clusters, some 25 miles north-east of the town of Cartwright on the Labrador coast. I spent three exhilarating summers in the early 1980s at this, one of Canada's most diverse seabird colonies. Five of the North Atlantic's six auk species breed here each summer: common guillemot, Brünnich's guillemot, Atlantic puffin, razorbill and black guillemot. One day a little auk – a species that breeds much further north in Greenland – visited too, perching momentarily beside our camp. The numbers of seabirds are incredible: tens of thousands of common guillemots and puffins, fifteen thousand pairs of razorbills, and smaller numbers of the other two species. There are no gannets; indeed, the closest gannet colony is 300 miles to the south on Funk Island, making the name of these islands a mystery.[1]

During the occasional interludes when I needed a few minutes alone, away from my five seabird-studying colleagues with whom I shared the islands and our cramped accommodation, I came to this remote spot and wondered what it would be like to discover a colony of great auks unknown to the rest of the world.

The focus of my fantasy was their biology: how they lived and bred, and how their lives would differ from their extant relatives. What would it be like to study them? As far as we know, great auks never bred on the Gannet Islands, but they certainly occurred in other parts of Labrador in the distant past.[2]

What never really occurred to me during my great auk daydreams was the ethical dilemma such a discovery would generate. This is the theme of Michael McCarthy's wonderful book *Fergus the Silent*, published in 2021. The question is this: if you were to make such an earth-shattering find, would you tell the world? Would you risk the birds being captured and brought into a captive-breeding programme, or would you remain silent like the hero of McCarthy's book?[3]

The official extinction date of the great auk is 3 June 1844 – the actual day in June is uncertain, as is the year, so the date is symbolic. Whichever day it was, the last two specimens known were killed on the island of Eldey and their skins and internal organs taken 'in the name of science'. The truth is that they were taken simply for greed, and it was sheer luck that their internal organs were preserved. Despite the great auk being declared extinct, there were some subsequent sightings after 1844.

Knowing what we now know, this was not unexpected. The great auk was naturally long-lived, with some individuals perhaps attaining the age of 50. A common guillemot chick I ringed on the Gannet Clusters in Labrador in 1981 was found dead 40 years later. A razorbill hatched and ringed on Bardsey (the Welsh island Hewitt once thought of buying) was seen alive and breeding 41 years later. Since large birds tend to live longer than smaller ones, 50 does not seem unreasonable as the maximum age of a great auk. Razorbills first start to breed at between four and six years old, and common guillemots not until they are six or seven. Great auks must have been similar. During their first few years of immaturity, razorbills and guillemots spend most of their time out at sea away from the colony. Only as the urge to breed starts to stir do they visit the breeding sites. Immature guillemots, three, four or five years old, often congregate in their 'clubs' on tidal rocks at the foot of the cliffs, or in areas adjacent to breeding areas. These young birds are nervous and leave at the slightest disturbance, such as a gull's alarm call, a swooping falcon or the sight of an approaching human.

All of this means that when the last two breeding adults were killed on Eldey, any immature birds visiting the colony on that day in June would have left long before Vilhjálmur Hákonarson or any of his crew set foot on that fateful shore. Other birds hatched in previous seasons and not yet old enough to breed would have been somewhere out at sea.

This assumes that there were young great auks to have been 'invisible' at sea. As we saw earlier, the men seeking adult great auks and their eggs invariably killed everything they could catch and either 'trod down' or collected any eggs they came across. That, plus the fact that no chicks were ever seen (or collected) from Eldey raises the question of whether in the last few years prior to 1844 *any* great auk chicks were hatched and reared. It is perfectly possible that some were, because there were no visits to Eldey between 1835 and 1839, nor in 1843, so some of the few remaining pairs may have reared a chick in those years. Perhaps it was some of these birds that were subsequently reported.

In 1848, *four* great auks were reported off Vardo, in Varanger Fjord, Norway, one of which was shot and killed. The bloody corpse was left on the beach, but when the hunter, a Herr Brodtkorb, returned to collect it the next day, it had disappeared. Vardo is too far north to be entirely plausible, and Alfred Newton, who quizzed Brodtkorb about the sighting, wasn't convinced. However, Newton's friend John Wolley was.[4]

In December 1852, Colonel Henry Maurice Drummond-Hay, one of the founders and an early president of the British Ornithologists' Union, was sailing from Halifax, Nova Scotia to Europe. As he passed 'over the tail of the Newfoundland banks' he saw what he fully believed to be a great auk within 30 to 40 yards (around 30 metres) of his boat. Drummond-Hay wrote to his friend Alfred Newton to tell him of the sighting, saying that he 'could see the large bill and white patches [on the head], which left no doubt in his mind'.[5] I am sceptical, however. In December a great auk would probably have been in its winter plumage – a fact that Drummond-Hay failed to mention. We know from the two winter-plumage specimens that exist (see p. 108) that the bird's white 'spectacles' in front of the eye are either absent or indistinct at this time of year. Despite his impeccable credentials, Drummond-Hay may have been mistaken.[6]

In 1853 a great auk was found dead in Trinity Bay on the east coast of Newfoundland. Three years later, in 1856, another one was said to have been caught on the island's west coast. There were further records, including a sighting of two birds in 1869 by an Icelandic mariner,[7] but the more the interval since 1844 increased, the less plausible they seemed.

One seemingly authentic but very late record came from the Lofoten Islands, Norway, in the 1930s. The great auks that people claimed to have seen there were in fact king penguins, nine of which had been transported from the southern hemisphere and released in 1936 off northern Norway. This was one of many misguided attempts to introduce species to areas where they did not naturally occur: reindeer shipped from Norway to South Georgia; musk oxen from Greenland to Svalbard; Arctic foxes from Alaska to the Aleutian Islands. All of these events were successful. Unfortunately for the penguins in Norway but fortunately for us, they did not establish a population, and the last confirmed sighting of one was in 1948.[8]

None of the post-1844 great auk records can be verified. Ornithologists keen to believe in the existence of an 'extinct' species seem to be easily deluded. The ivory-billed woodpecker, an impressively powerful bird of the southern United States, was first described in 1731. Logging destroyed its habitat, causing numbers to crash. Scientists played a role, collecting no fewer than 400 specimens. The last birds were seen in 1944 – a century after the last great auks – when the woodpecker was declared extinct. But as with the great auk, there were later purported sightings, right up to 2021, but no unambiguous confirmation. It is difficult, of course, to confirm a negative. Encounters after an official extinction date reflect both a desperate need for a species not to have gone for ever, but also, possibly, a bid for fame.[9]

There is, however, a handful of examples of birds once assumed to be extinct that really did come back to life. The exquisite, blue-bearded helmetcrest hummingbird had not been seen for 69 years when it was rediscovered in Colombia in 2015. The blue-eyed ground dove of Brazil, absent and assumed extinct for 75 years, was also seen in 2015, while the black-browed babbler had been lost for 172 years when, in 2020, it was found again in Borneo.

The difference between these three thought-extinct species and the great auk is that the hummingbird, dove and babbler are all birds of remote and dense forest habitat in which it is much easier to remain undetected. The great auk bred out in the open and wintered at sea, making it much less likely that it could remain undetected for 180 years.

Unquestionably, it has gone. Yet I feel, too, that the great auk has an enduring presence.

The archaeological bony remnants of great auks are an important part of the birds' afterlife. Apart from the DNA they contain – of which more below – these remains tell us more about ourselves than the biology of the bird itself. In the 1970s, when my own auk studies were just beginning, I read of the discovery in Newfoundland of an ancient burial site. Poignantly for me (and the archaeologists), some of the graves contained the beaks and gizzard stones from great auks (see p. 43). Four thousand years ago, these people clearly held the great auk in high esteem. Two millennia later, and quite independently, the Egyptians were also interring birds or bits of birds with human bodies. For them, and probably for the indigenous people of Newfoundland as well, birds were the enduring link between this life and the next – the afterlife. Winged messengers, connecting earth with heaven, or more appositely in the auk's case, with the ocean. We might smile at such archaic ideas, but for several centuries, cherub-bird hybrids – otherwise known as angels – have served the same purpose for Christians.

Not only did the Egyptians embalm and preserve the bodies of ibises and falcons, they also created images of different birds on the walls of their tombs, either in paint or carved in stone. As I mentioned earlier (see p. 40), there are also several palaeolithic caves containing images of great auks. Such enduring visual images *are* the afterlife of both the birds and the people they represent.

In certain religions, the afterlife entails some kind of judgement: a life well-lived? Devoid of souls, according to some at least, non-humans were excused judgement. Certainly, if great auks were to be judged, we would find their souls to be spotless. Not so their mass-murderers. Judged now, we condemn and despair of those who caused such carnage at Funk Island and the extinction in Iceland.

The most important, obvious and tangible manifestations of the great auk's afterlife are its mortal remains: 75 empty eggshells, two sets of spirit-preserved internal organs, skeletons, individual bones, and the 80 or so mounted specimens. There exist, as we have seen, replica eggs and fake specimens, too. In addition, there are innumerable images and some sculptures, but they count only as imperfect reminders of what we once had. Important, certainly, but – depending on who created them – there's only so much information you can extract from a painting or a photograph. An example of a scientifically valuable image is the 1844 painting by Johann Naumann of 'Benicken's Winter Auk' painted, if not from life, then accurately from its skin (see p. 109). Edward Bidwell's photographs (albeit black and white) taken in the 1890s of 69 great auk eggs are also extremely useful.

I recounted earlier the value of real relics in rectifying previous errors, such as the number of brood patches the great auk possessed (see p. 110). In a similar way, genuine great auk eggshells have allowed us to quantify the shape of their eggs and to use that information to assess and infer different aspects of the bird's breeding biology.

After decades as an ardent egg-collector, Alfred Newton, professor of Comparative Anatomy at the University of Cambridge and Britain's most eminent ornithologist, was forced to conclude in 1896 that oology had little scientific value. His statement was rooted in the fact that the central question taxing ornithologists at that time was the arrangement

– or phylogeny – of birds. Here is Newton, in all his pompous Victorian expansiveness:

> *The present writer ... must confess to a certain amount of disappointment as to the benefits it* [oology] *was expected to confer on Systematic Ornithology, though he yields to none in his high estimate of its utility in acquainting the learner with the most interesting details of bird-life – without a knowledge of which nearly all systematic study is but work that may as well be done in a library, a museum, or a dissecting room, and is incapable of conveying information to the learner concerning the why and the wherefore of such or such modifications and adaptations of structure.*[10]

Eggs, Newton reluctantly admits, were no better than tongues, brains, palettes or guts as reliable clues to the phylogeny of birds. Oology's main motivations, he admitted, remained aesthetic, hunting and status-enhancement.

Newton's statement, together with general concern over the killing of birds for feathers or for 'sport', nevertheless helped to launch the anti-oology movement in Britain, one of whose most vociferous and articulate proponents was Max Nicholson. Precociously capable, Nicholson wrote a series of ground-breaking bird books in the 1920s and 1930s. As well as serving to encourage many new areas of ornithological science, he also promoted the conservation of birds. Concerned by the damage inflicted by collectors, taxidermists and oologists, Nicholson said 'directly or indirectly, collecting [eggs and skins] is the root of almost all calculated evil for our bird-life'.[11]

Decades of campaigning led eventually to the 1954 Protection of Birds Act that made the collecting of wild birds' eggs illegal. Part of the Act's justification was to protect bird populations, but it was also because the oologist's mask of scientific respectability had been ripped away. Nicholson was especially irritated that collectors had singularly failed to address some of the most fundamental biological questions about birds' eggs, such as why their relative size, surface texture and colour varied so much between species – omissions that scientists rather than oologists have rectified only in the last decade or so.[12]

The 1954 Act stopped most oologists in their tracks, but it had another consequence. It made the biological or cultural study of eggs a forbidden, or at least questionable, area of ornithology. Despite the essential role of museum eggs in identifying the disastrous consequences of DDT in the 1960s and acid rain in the 1990s, it still took a full 50 years for the heat to die down. Then, in the early 2000s, there grew an increasing awareness that the large collections of eggs – including some once owned by Vivian Hewitt – held in national museums had huge potential, both scientifically and culturally.[13]

The change that occurred is readily apparent from various studies, including some on the great auk, whose eggs and skins have provided us with a new and unexpected window into its way of life. This means that the idea – falsely rooted in the now-outdated notion that they have no scientific value – of destroying old collections, as advocated by bird protection organisations, is misguided. While we certainly would not want to see a resurgence in unregulated egg-collecting, destroying eggs already collected is completely pointless.

It is the fact that eggs and skins *are* biologically informative that has driven my obsession with finding Hewitt's missing eggs. Knowing where great auk remains are housed is important because, in some instances, having access to them may forge that all-important link between history and science, and enrich the development of both.

The great auk's least visible but arguably most important relic is its DNA. Extracted from bones, skin, the heart and other internal organs, and potentially from the membranes inside its eggshells, DNA has proved crucial for establishing various aspects of this bird's biology. Among the first molecular results was the unambiguous confirmation of the great auk's phylogenetic relationship with other auks. In turn, this provided biologists with a robust framework for making comparisons between the great auk and its closest relatives, the razorbill and the two guillemot species.

In Chapter 3, I mentioned Sven-Axel Bengtson's suggestion that the great auk's demise was a consequence of climate change in the Middle Ages. When he was writing those words in 1984, the molecular revolution had barely begun. Like the rest of us at that time, Bengtson could hardly have imagined the technological innovations that would eventually provide the tools needed to answer his question.

It took more than 30 years, but by 2019 we had an answer. It is a story that starts with John Stewart at Bournemouth University in the UK. Stewart had long been fascinated by fossil auks. He had actually discovered a new species of extinct auk in 2000 that, thanks to a colleague, bears his name, *Alca stewarti*. It was a cousin of the great auk and probably similar in design to the razorbill, but at an estimated two kilograms it was twice as heavy.[14]

In 2010, Stewart and his colleague Michael Hofreiter applied to the UK's Natural Environment Research Council for a grant to investigate the history of the great auk using 'ancient DNA', that is, DNA extracted from its surviving relics. Their application was unsuccessful – as most are because funds are so tight. Undeterred, Stewart and Hofreiter found another opportunity to pursue the same question by collaborating with a newly appointed colleague, Michael Knapp, and a different source of funding – a PhD studentship.

Originally, the project was intended as an investigation of the impact of environmental change on seal numbers. After failing to secure the great auk grant, however, it was suggested to the recently appointed PhD student, Jessica Thomas, that she switch from seals to the great auk. The approach was similar – using molecular methods to analyse DNA extracted from bones – to establish whether the great auk became extinct as a result of exploitation or because it was already in decline for reasons such as climate change, as Bengtson suggested.

Thomas used mitochondrial DNA extracted from the humeri of 41 different great auks. These bone specimens, recovered from various archaeological sites, ranged from 175 to 15,000 years old. Together with some clever mathematical modelling, the material allowed Thomas and her colleagues to reconstruct the great auks' population structure and population dynamics. The rationale was this: if environmental change had been responsible for the great auk's demise, this would be reflected by a reduction in genetic diversity over time. On the other hand, if persecution was the cause, no such reduction would be apparent.

The results were clear. There was no trace of the genetic signature indicating a population already in trouble from environmental change. Instead, the great auk's extinction was the result of nothing more than human need and greed.[15]

In an old-fashioned bottle of spirits in the Natural History Museum of Denmark rests what looks uncannily like a miniature human heart. It is a heart that once beat in the breast of a great auk. It is also the spirit of a great auk. It is there because on that June day in 1844, when this bird and its partner were so brutally killed and skinned, their hearts, eyes, gizzards, gonads and other organs were removed and plunged into 'spirit'.

We do not know which of the 70 mounted skins are the last two great auks; we have hearts but no bodies. Identifying them became one of Errol Fuller's goals when writing his great book in the 1990s. He narrowed the options down to five candidates: the Bremen, Brussels, Kiel, Oldenburg and Los Angeles birds. The latter two were considered the most likely because these were thought to be the two birds handled by the dealer, Israel of Copenhagen, in 1844. Using mitochondrial DNA extracted from the skins of these five birds, and from the heart of the female and the oesophagus of the male spirit-preserved birds in Copenhagen, Thomas and her colleagues looked for matches. After many laborious hours in the lab, there was just one match, rather than the two they had hoped for. The mismatches between the DNA from the skins and organs of the Bremen, Kiel and Oldenburg Auks discounted all of them. There was a good match, however, between the male's oesophagus DNA and the Brussels Auk skin. Thus it was confirmed that the Brussels Auk was one of the final two birds to have been seen alive.[16]

Contrary to Fuller's expectation, the match between the female's heart DNA and the Los Angeles skin was poor, and sufficient to discount it. But why? If Israel of Copenhagen's second auk – that is, the Los Angeles Auk – wasn't one of the two last birds, which one was?

As we saw in Chapter 4, before being bought by Vivian Hewitt in 1934, both the Cincinnati Auk and the Los Angeles Auk had belonged to George Dawson Rowley. Perhaps the two birds were simply mixed up at some point. Maybe the bird Rowley thought was the Los Angeles bird was actually the Cincinnati Auk, and *vice versa*. After all, the two specimens look remarkably similar (see p. 152).

To check meant going back to the lab for Jessica Thomas. I was birdwatching at Hickling Broad in Norfolk on the day Errol Fuller went to visit Thomas in her new position at the

University of Swansea. She told him the news, and he phoned me to pass on her discovery, confirming Fuller's suspicion that the other 'last' auk *was* the Cincinnati specimen. The two had indeed been mixed up.[17]

We now know who those skins and organs belong to. It is akin to returning anthropological artefacts to their country of origin, or reuniting bits of disinterred bodies.

Advances in molecular biology have also raised the possibility of de-extinction: the recreation of long-dead species from fragments of their DNA that still survive. It is an alluring prospect, and who wouldn't want to see a living mammoth, dodo or great auk? I certainly would, and, initially at least, I was enthusiastic about the possibility. But then as I thought about it, my enthusiasm was supplanted by some cold, hard facts. No less than half the world's seabirds – over 200 million individuals – have been lost since the 1950s. The causes comprise a shameful anthology of anthropogenic effects: oil and other chemical pollutants, over-fishing, introduced predators, by-catch in gill nets, disturbance, hunting, climate change-driven changes in food availability, and most recently, avian flu. Each of these anthropogenic effects has had a different trajectory and temporal profile. The sickening effects of oil pollution, for example, were most prevalent in the twentieth century; avian flu was apparent only in the twenty-first. Had great auks not been extirpated by hunting, could they possibly have survived the subsequent barrage of anthropogenic killing-agents?[18]

In a word – no. Being flightless, with a much-reduced population and breeding at just a handful of sites, the great auk would have been desperately vulnerable to any or all of these threats. The current climate-change shift in fish distribution would have been particularly devastating for a flightless bird, just as Bengtson previously speculated. Since the 1970s, puffins and guillemots in the north-east Atlantic have suffered from food shortages during the breeding season. The first indications of this were mass die-offs of chicks in Norway's once-huge puffin colonies. Guillemots have experienced similar issues, resulting in birds either failing to breed at all or, like puffins, having to watch their chicks starve to death. In south-east Iceland, the situation is similar, with marked declines in auk breeding success and numbers. These disastrous effects are not (yet) apparent in the north-west Atlantic, including in seabird colonies like those on Funk Island where great auks once bred. If great auks were alive today, those on Funk Island might have escaped the climate-change effects – assuming they had survived the other threats, including the mass over-exploitation of commercial fish in the area.[19]

If the current environment of the North Atlantic is no longer suitable, then we should be wary of trying to resurrect the great auk. The idea of de-extinction is both seductive and deceptive. It sounds like confession: commit whatever sins you like, including extinction, then confess and all is forgiven and restored. Sorry, that just doesn't wash. But it's seductive, yes. The possibility that through molecular wizardry we could recreate the great auk and once again enjoy the sight, sound and smell of it I find almost irresistible.

There is a long way to go, both technically and ethically. Even if we could recreate the great auk, would it be fair to spend all that time and effort doing so for a bird for whom there is probably no safe home? Wouldn't it be better to spend those funds saving one or more endangered species not yet extinct, and whose habitat could be made safe?[20]

Epilogue:
A Less Perishable Inheritance

In the 1860s, Alfred Newton wondered whether any great auks still existed, and worried that even if they did, they were desperately vulnerable. Hoping that a few still survived, he wrote: 'I ... beseech all who may be connected with the matter to do their utmost that such a rediscovery should be turned to good account. If in this point we neglect our opportunities, future naturalists will reproach us.' Indeed.

Extinct means gone for ever. Even if molecular technology were to eventually allow us to recreate the great auk, its extinction in 1844 marked a turning point in our relationship with these birds. A turning point full of contradictions. On the one hand, I despair of how the greedy obsession for acquiring 'specimens' eventually drove the great auk to extinction. On the other, I am relieved that those specimens – eggs, skins and so on – still exist, because they are such poignant reminders of what we have lost. It is easy to say that the way men behaved in the early 1800s was another era with different values. It was, but even in those few cases where someone, like George Cartwright, urged restraint or concern, they were silenced by the overwhelming need to acquire.

Alfred Newton, one of my ornithological heroes, was one of those who experienced such contradictory sentiments. Even he could not help himself. *Killing first for specimens, and then observing living birds only if there were sufficient skins.* Those were Newton's written instructions for an assistant whom he sent to another potential great auk colony in Iceland. 'If there are great auks there, this is what you will do' – the very same rules that he and John Wolley intended to apply themselves had they found great auks on Eldey.[1]

Despite my sense of moral outrage at the loss of the great auk, I can also understand Newton's and Wolley's motivation. We are all trapped, one way or another, within certain sets of beliefs and values that encourage, if not compel, us to act in certain ways. Only when there's a paradigm shift do we suddenly reflect on what might have been. The change in UK law in 1954 that made collecting birds' eggs illegal was one such paradigm shift. An activity that was once legal, respectable and even educational was transformed into something shameful and felonious.

In one sense, making egg-collecting unlawful was an easy way to trigger change. Unfortunately, we cannot make it unlawful to allow a species to go extinct. What we can do, though, is to reflect on those cases we know about: the dodo, the passenger pigeon, the ivory-billed woodpecker, the huia, the Stephens Island wren and the great auk. We can then decide what we should draw from these examples.

Alfred Newton's obsession with the great auk led to his only significant biological discovery: that humans – *Homo sapiens*, that is, 'wise man' – could cause a species to go extinct. Up until that point, extinction had been assumed to be a consequence of natural events: a cooling climate rendering life unsupportable for dinosaurs; one species outcompeting another for food, and so on. Following their visit to Iceland, Newton and Wolley's realisation that humans were responsible for the loss of an entire species hit them like a hammer blow. Wolley's own extinction – his premature death in 1859 – dealt a similar blow to Newton, and together those two events precipitated a shift in the way he viewed the world. It triggered a passion for protecting birds. That passion eventually grew into the bird protection movements we see today.

And it has seen some notable successes, with a handful of birds brought back from the brink of extinction. By 1980, New Zealand's black robin was down to just five individuals, of which only one was a breeding female, but careful management in the wild saw the population increase to reach more than 300 individuals by 2021.[2] The California condor's population was reduced – through pesticides, poaching, and lead poisoning – to just 27 individuals in 1987 when a hugely controversial decision was taken to capture all the survivors and create a captive breeding programme. It has been a success, and by 2022 the world population had risen to over 500 individuals.

In some ways, the bird whose plight most closely matches that of the great auk is New Zealand's flightless parrot, the kakapo. Like the great auk, it is the largest of its kind, and like the great auk it was abundant and widespread and played an important role in the culture of the indigenous people. The arrival of Europeans and their dogs plus other predators, such as Polynesian rats, from the mid-1800s caused a collapse of the kakapo population. As with the great auk, the scarcer the kakapo became, the more desperate scientists became to obtain 'specimens'. Thousands were collected and killed for zoos and museum collections. Conservation efforts started early, in the 1890s, but it was a huge challenge and numbers continued to decline. The bird came within a filoplume of extinction. It was not until 1995, when numbers had fallen to just 50 individuals, that the Kakapo Recovery Programme was established. By creating the equivalent of an intensive care system *in the wild*, numbers started to increase. By 2022 the world population stood at 252 individuals.[3] Seeing kakapo on Codfish Island in 2012 was, for me, the antipodean equivalent of having seen a great auk.

Nonetheless, these measures can sometimes seem little more than a finger in the proverbial dyke. Around that finger, birds everywhere are leaking away, their numbers declining year on year.

We must not underestimate the magnitude of what is required to halt or reverse these declines. But it is a challenge we must rise to and one that requires big thinking, rather than fiddling while Rome burns, as we seem to be doing at present. We ourselves are being pitched towards an uncomfortable afterlife; something many of us can see but feel helpless to stop. As Newton said: 'The mere possession of a few skins or eggs, more or less, is nothing. Our science demands something else – that we shall transmit to posterity a less perishable inheritance'.

Let us celebrate the magnificent great auk and what its remains can continue to teach us about the value of life.

Appendix 1

Putative Great Auk Breeding Colonies

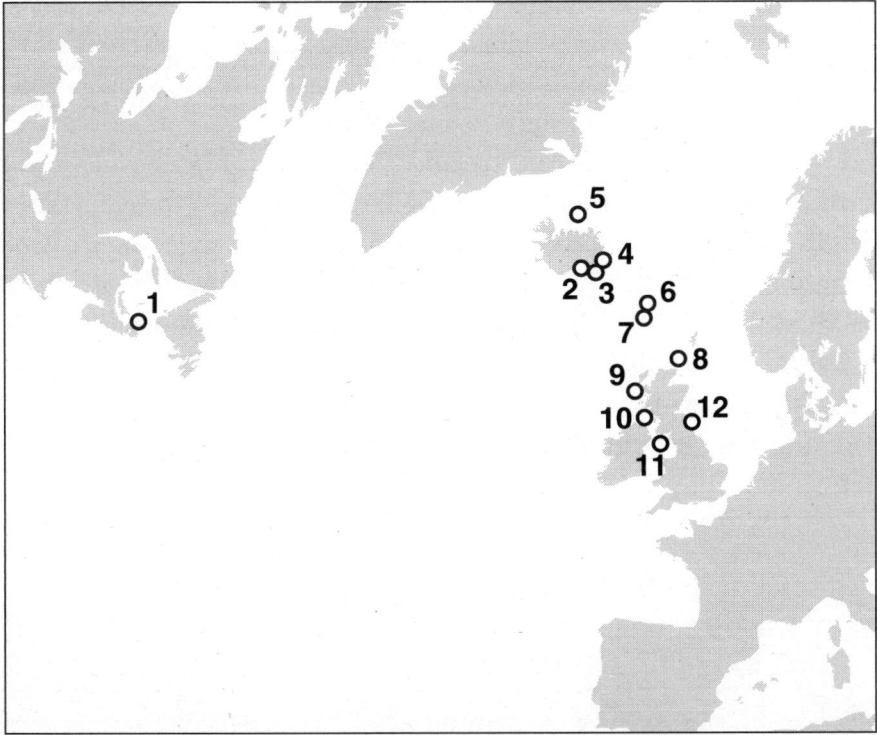

Cape Breton Island, Nova Scotia (1), Tvísker, southeast Iceland (2), Hrollaugseyjar, south-east Iceland (3), Hvalbakur, south-east Iceland (4), Kolbeinsey, northern Iceland (5), Fugloy, Faroe Islands (6), Streymoy, Faroe Islands (7), Tofts Ness, Sanday, Orkney Islands (8), Oronsay, off Colonsay, western Scotland (9), Bòrnais, Uist, western Scotland (10), Calf of Man, Irish Sea (11), and the Farne Islands, north-east England (12).

Nettleship & Evans (1985: table 2.1, p. 64) list nine putative colonies, but point out that three of these were extremely unlikely: the Penguin Islands of north-east Newfoundland, the Penguin Islands to the south of Newfoundland and Gunnbjörn's Skerries in south-east Greenland. I have therefore excluded these. I have included all others on their list – Cape Breton Islands, Hvalbakur, Tvísker, Fugloy, Streymoy and Calf of Man.

The putative record of great auks breeding on Lundy Island in the Bristol Channel, England (see Serjeantson 2023, citing Baldwin 2009) refers simply to a very large, pointed egg, which could have just as well have been a double-yolked guillemot egg (see Birkhead *et al.*, 2021). I have discounted Lundy.

I have added to the list of putative colonies:

Hrollaugseyjar, south-east Iceland (Steenstrup 1885), Kolbeinsey, northern Iceland (Kålund 1879–1882), Tofts Ness, Sanday, Orkney Islands (Serjeantson 2023), Oronsay, off Colonsay, western Scotland (Serjeantson 2023), Bórnais, Uist, western Scotland (Serjeantson 2023), and the Farne Islands, north-east England (Jackson *et al.* 2022).

I also checked that there had been no changes since Nettleship & Evans's paper to putative colonies in either the Faroe Islands (there had not; J-K. Jensen, pers. comm.) and Greenland (M. Meldgaard, pers. comm.).

Appendix 2

Vivian Hewitt's Thirteen Great Auk Eggs

	1	2	3	4	5	6
Date of Hewitt purchase	1934	1934	1935	1937	1939	1939
Fuller's name	Dawson Rowley's	Lord Garvagh's	Lady Cust's	Vivian Hewitt's	Cincinnatti	Wallace Hewitt's
Fuller's number	18	19	21	12	66	70

TIME ↑	F Rowley	F Rowley	Lupton 1935 Jourdain 1934 F Rowley	Son of Vaucher	Massey 1889	
		G. D. Rowley c.1870	G. D. Rowley 1878	Vaucher, Lausanne 1900	L. Field 1867	Massey 1894 via Stevens's Auction Room
	J. Gould to G.D. Rowley 1865 Stevens's Sales Room 1865			de Méezmaker date?	Burney 1865 via Stevens's Auction Rooms	
			Bought by Yarrell for Lady Cust c. 1878?	Merchant in Bergues, France		Wallace Hewett 1894
	Hunterian Museum 1861				Hunterian Museum 1861	
		Lord Garvagh 1853 Stevens's 1852 Potts c. 1852/3 Gardiner 1852				
				Captain of whaling ship pre-1800		
		Boulogne Museum 1825				
		Vicompte de Barde pre-1795				
	G. Cartwright to Hunter 1773?				G. Cartwright to Hunter 1773?	

7	8	9	10	11	12	13
1940	**1940**	**1946**	**1946**	**1946**	**1947**	**1948**
Bowman Labrey's	Yarrell's	Dr Dick's	Alfred Newton's	Jack Gibson's	Captain Cook's	Malcolm's
20	22	16	15	17	73	72
F Rowley	Lupton 1934				F Rowley	
G. D. Rowley	Hirsch via Stevens's Auction Rooms 1925	Hunterian Museum 1861	Hunterian Museum 1861	Hunterian Museum 1861	G. D. Rowley 1863	
Bowman Labrey 1871					Gould 1863 from German Bazaar, London	
Wilmott or Yarrell c. 1844	Vauncey Crewe 1894 via Stevens' Auction Rooms					
Leadbeater 1840s						
	Louis d' Hamonville 1875					Leadbeater 1840 to Malcolm
	Gardiner 1856 via Stevens' Auction Rooms					
	Yarrell 1815					
		G. Cartwright to Hunter 1773?	G. Cartwright to Hunter 1773?	G. Cartwright to Hunter 1773?		

Appendix 3

List of Bird Species Mentioned in the Text

Alpine accentor *Prunella collaris*
Ancient murrelet *Synthliboramphus antiquus*
Atlantic puffin *Fratercula arctica*
Barn swallow *Hirundo rustica*
Black guillemot *Cepphus grille*
Black-browed babbler *Malacocincla perspicillata*
Blue-bearded helmetcrest hummingbird *Oxypogon cyanolaemus*
Blue-eyed ground dove *Columbina cyanopis*
Brünnich's guillemot *Uria lomvia*
California Condor *Gymnogyps californianus*
Common cuckoo *Cuculus canorus*
Common guillemot *Uria aalge*
Dodo *Raphus cucullatus*
Elephant bird *Aepyornis* spp.
Eurasian goldfinch *Carduelis carduelis*
Gannet (or northern gannet) *Morus bassanus*
Great auk *Pinguinus impennis*
Great black-backed gull *Larus marinus*
Greenshank *Tringa nebularia*
Herring gull *Larus argentatus*
Hoopoe *Upupa epops*
Huia *Heteralocha acutirostris*
Ivory-billed woodpecker *Campephilus principalis*
Kakapo *Strigops habroptila*
Kaua'i 'Ō'ō *Moho braccatus*
King penguin *Aptenodytes patagonicus*
Labrador duck *Camptorhynchus labradorius*
Lammergeier (bearded vulture) *Gypaetus barbatus*
Lapwing – see Northern lapwing
Little auk *Alle alle*
Little Blue Penguin *Eudyptula novaehollandiae*

Meadow pipit *Anthus pratensis*
New Zealand black robin *Petroica traversi*
Northern gannet *Morus bassanus*
Northern lapwing *Vanellus vanellus*
Osprey *Pandion haliaetus*
Passenger pigeon *Ectopistes migratorius*
Peregrine falcon *Falco peregrinus*
Purple sandpiper *Calidris maritima*
Raven *Corvus corax*
Razorbill *Alca torda*
Red kite *Milvus milvus*
Red-backed shrike *Lanius collurio*
Rodrigues solitaire *Pezophaps solitaria*
Slavonian (or horned) grebe *Podiceps auritus*
Spix's macaw *Cyanopsitta spixii*
Stephens Island Wren *Traversia lyalli*
Turnstone (or ruddy turnstone) *Arenaria interpres*
White stork *Ciconia Ciconia*

Acknowledgements

The great auk and the millionaire Vivian Hewitt have both fascinated people for many years. Of the two the auk is the better known, although for both of them I have had to sift through vast quantities of remains to piece together their interdependent story.

My greatest debt is to David Clugston, the birder and bibliophile who saved all of Vivian Hewitt's papers and correspondence after David Wilson's death and made them available to me. Without his help there would have been no story. Others have been intrigued by the mysterious life of Captain Vivian Hewitt. These include William Hywel Jones, Hewitt's doctor, whose biography *Modest Millionaire* (1973), written under the name of William Hywel, has been an invaluable source of information. After Hewitt's death in 1965 it was David Wilson of the British Trust for Ornithology who ensured that all of Hewitt's papers were saved, and which he subsequently stored in his home. Despite his best efforts, Wilson was less successful at salvaging Hewitt's collections. There was simply more material than anyone could reasonably manage. Indeed, during his lifetime Hewitt failed to manage his own collection. I was amazed to discover that 50 years after Hewitt's death, Wilson, long since retired, was still visiting Hewitt's home in Anglesey and trying to rescue some of the remaining artefacts. David Wilson died in 2020. A relative, knowing that David Clugston had been a friend, asked him to sort through Wilson's houseful of books, papers and specimens to decide what was worth keeping. Had it not been for David Clugston and Steve Holliday, the several suitcases of correspondence and much else besides would have undoubtedly ended up in a skip. David Clugston subsequently read and catalogued much of Hewitt's correspondence, and proved to be a selfless and perceptive reader off whom I could bounce ideas and pester with queries.

The second person to whom I am particularly indebted is Errol Fuller, author of *The Great Auk* (1999), the definitive account of the species. We had known of each other and each other's interest in the great auk since the 1990s, but it was only when David Clugston was helping to empty David Wilson's house that we finally made contact and met, and for that I also thank Jim Whitaker. David Wilson had helped Fuller with Hewitt's correspondence when he was writing his great auk book. Errol knows more about the great auk than anyone else, and as his book testifies, there is a vast amount to know. I have been extremely fortunate to have been able to seek his advice on numerous occasions.

I am also extremely grateful to my academic colleague Bob Montgomerie of Queens University, Ontario. For 30 years we have shared an interest not only in scientific ornithology but also in its history. As I wrote, I asked him for his views on great auks and other topics, knowing that I would always receive a sensible, honest and often inspirational response.

Another colleague with whom I have collaborated for many years is my medic friend, Karl Schulze-Hagen. Some of the early great auk literature is written in German, and I have benefited enormously from the fact that Karl became as intrigued as myself with the great auk. Karl delved into the great auk's German history with enthusiasm and clinical precision, and made some extraordinary discoveries. It was through Karl that I was able to visit the Naumann Museum in Köthen, Germany, and meet both Jürgen Fiebig, curator at Berlin's Natural History Museum, and Klaus Nigge, a professional photographer, both of whom have played a crucial role in our great auk research.

Michael McCarthy, one-time environmental journalist for the *Independent* newspaper, was also a source of inspiration as he wrote his novel *Fergus the Silent* (2022), at the heart of which lies the great auk.

In Iceland, where the last great auks died, three colleagues have been especially helpful. Gísli Pálsson, whose own book on the great auk has been a novel source of inspiration, has been selfless in his help. Thanks also to Arni Snaevaar and especially to my colleague from our PhD days, Aevar Petersen, for sending me material that would otherwise have been inaccessible or unintelligible.

I received invaluable help from many others, too, including: Graham Axon, Nigel Brown, Laurent Chevrier, Glen Chilton, Ann Farrell (daughter of William Hywel Jones), Jon Fjeldså, Thord Fransson, Sarah Glassford, Dan Gordon, Christophe Gouraud, Nigel Harcourt-Brown, Mike Harris, Andrea Hart, Lesley Hindley (BTO Archivist), Matthew Holley, Helen James, Colin Jones (Rhyl History Society), Evan Jones, Martyn Linnie, Bruce Lyon, Bob McGowan, Richard Mearns, Morten Meldgaard, Chris Milensky, Wendy Moore, Charles Nelson, David Nettleship, Sean Nixon, Brian Oliver, Ian Owens, Verity Petersen, Jonathan and Jane Price, Gloria Ramello, Matt Ridley, Ifan Roberts (Anglesey Archives), Douglas Russell, Nicola Samuel, Dale Serjeantson, Guy Shorrock, Mike Siva-Jothy, John Stewart, Frank Sulloway, Paul Sweet, Ann Sylph, Jessica Thomas, Jamie Thompson, Julian Thompson, Jennifer Vess, John Walters, Sarah Wanless, Jan Whitaker, Jim Whitaker, John Wiens, Sabine Wilhelm, Llew Williams and Bernie Zonfrillo.

Mike Birkhead, David Clugston, Bob Montgomerie and Jeremy Mynott all read various drafts and I am extremely grateful for their comments and insights.

Finally, I thank my agent Carrie Plitt at Felicity Bryan Literary Agency, and Jim Martin, my publisher at Bloomsbury, for their unstinting help and encouragement.

Notes

Prologue

1. Graham Axon's (b. 1958) extraordinary ability was not unique. A number of oologists in the past were able to complete these three-dimensional eggshell jigsaw puzzles. One was Lieutenant-Major Michael Prynne (1896–1972), an egg-collector whose party-piece was to take a chicken egg, smash it with his hand and then reconstruct it in such a way that no one could tell that the shell had been broken. That curious talent saw him invited in 1960 onto the BBC television show *What's My Line*, where he defeated the panel, who failed to guess his 'occupation' (Prynne 1963). Prynne's and Axon's ability to repair eggshells requires enormous skill and patience, but is facilitated by the way birds' eggshells are constructed. Unlike the fragments of a broken pot, the calcium carbonate crystals of a bird's eggshell are formed like the wedge-shaped capstone in a bridge, allowing fragments to be placed seamlessly together.
2. Geoff Parker (b. 1944), Parker (2021).
3. Bill Bourne (1930–2021); see Jenkins & Dunnett (1978) for a profile of Bourne.
4. Errol Fuller (b. 1945), Fuller (1999): the misreading of Bowman Labrey's (1817–1882) name seems to have been due to Francis Jourdain (1934). William Yarrell (1784–1856).
5. Edward Bidwell (1845-1929) photographed all known great auk eggs in the 1890s – see Fuller (1999).
6. The narrative pause image used throughout this book is redrawn from images of two great auks from page 19 of *The English Pilot*. Initially published in 1716, this great work was used as a navigational aid throughout the 18th century, with the auks appearing in all editions. That the great auk was sufficiently numerous on the Grand Banks to be used as a reliable navigational aid tells us something significant about its abundance, distribution and distinctiveness in these times. I also like the fact that whoever created those illustrations in the 1700s did so in a rather naive way.
7. Vivian Hewitt (1888–1965). More details of Hewitt appear later in this book.

8. Pieter Bruegel the Elder's (1525/30–1569) 16th-century depiction of King Herod's order to kill all human infants under two, translated into a Flemish village where the massacre is conducted by occupying Spanish soldiers and German mercenaries. The painting was once owned by Emperor Rudolf II, who had some of the more grisly scenes painted over. Even so it is horrific. (https://en.wikipedia.org/wiki/Massacre_of_the_Innocents_(Bruegel))

Chapter 1: Funk Heaven

1. Wagner (1998) introduced the concept of a 'hidden lek' through his study of razorbills on Skomer.

2. Great auk body mass: ornithologists generally refer to the size of a bird in terms of its weight (i.e. body mass). Remarkably, given the thousands killed, no one, it seems, ever weighed a great auk. Even so, within the great auk literature a body mass of 4,500g or 5,000g (10lb or 11lb) is often quoted. The 4,500g value comes from Henry Wemyss Feilden (1838–1921), an army officer, explorer and natural historian who visited the Faroes in 1872. While there he visited an 80-year-old man who was said to be the last person on the Faroes to have seen a living great auk. In 1808 that man, with others, visited the island of Stóra Dimun and killed a single great auk. The old man said it weighed nine Danish pounds (9 x 494g = 4,447g) or about 4,500g. He may have weighed the bird, or it may have been a guess. Some later authors assumed a mass of 5,000g. Based on the relationship between skeletal features, egg volume and body mass in other auk species, my colleague Bob Montgomerie and I (in press) estimate that the great auk probably weighed 3,560g. White bill furrows: my personal observations of razorbills and great auk museum skulls.

3. *Guardian* article by Philip Hoare (2021).

4. Elliot *et al.* (2013).

5. Niko Tinbergen (1907–1988); David Lack (1910–1973); Richard Dawkins (b. 1941).

6. Moult – see Stresemann & Stresemann (1966). Dispersal from breeding colonies: this is based on the occurrence of great auk bones in archaeological sites. Note also that around 1000 BC, during

a cool period in southern North America, great auks occurred as far south as Florida (Brodkorb 1960).

7. Fleming (1785–1857): 'In winter, the brownish-black of the throat and fore-neck is replaced by white, as I had an opportunity of observing in a living bird, brought from St Kilda in 1822' (Fleming 1824).

8. Funk Island as an alcid idyll: the Beothuks lived in Newfoundland from about AD 500, and according to accounts from Europeans, including that of Joseph Banks (1743–1820) in 1766, visited Funk Island 'once or twice each year' to collect eggs and seabirds, presumably including great auks. No one knows when these visits started, but the indigenous population was small, and the disruption of the great auks' peace was probably very limited (Banks, cited in Lysaght 1971; Montevecchi & Tuck 1987).

Chapter 2: Funk Hell

1. Ganong (1964). There is no evidence that indigenous people visited Funk Island prior to arrival of the Beothuks before about 500 AD. João Álvares Fagundes (1460–1522); see Ganong (1964). Montevecchi & Tuck (1987: 48) point out that Gaspar de Corte-Real (1450–1501) may have landed on Funk Island in 1500.

2. Nettleship & Evans (1985).

3. Cartier (1491–1557); the quote is from Cook (1993), originally in French but here in English translation.

4. From Cook (1993).

5. Richard Hore (fl. 1536): see Montevecchi & Tuck (1987: 49). Notwithstanding their great auk feasting on Funk Island, Hore's party later ran out of food on the Newfoundland coast, where some starved to death and others were cannibalised (Morison 1971).

6. Thomas Pennant (1726–1798). His naming of the great auk: Pennant (1776). The great auk's original scientific name, established by Carl Linnaeus (1707–1778) in 1758, was *Alca impennis*, *Alca* meaning 'auk' and *impennis* meaning 'without flight feathers'. In 1791, Abbé Pierre Joseph Bonnaterre (1752–1804) proposed that the great auk deserved a genus of its own, which he called *Pinguinus*, apparently on the basis of the great auk's tiny wings. However, ornithologists continued to use both *Alca* and *Pinguinus* until the

Danish ornithologist Finn Salomonsen (1909–1983) in 1944 confirmed Bonnaterre's proposal, and *Pinguinus* became the standard (Salomonsen 1944; Olsson 1977). As Jeremy Gaskell (2000) has said, the alternative name in Castilian for 'el Pingüino' was 'el Pájaro Bobo', a derogatory adjective with the same meaning as 'Dodo', reflecting the great auk's (apparent) stupidity and the ease with which they could be caught and killed. Gaskell says that this reflects nothing more than their absence of fear, having existed without serious molestation on their breeding grounds for millennia. But there's more to it than that: sitting tight in the face of predators was their main means of defending their eggs and chicks from predatory birds, as is the case in common and Brünnich's guillemots and the razorbill. For the Beothuk name, see Gilbert (2011).

7.   Cook and Lane's 1775 map of Newfoundland is from here: https://www.heritage.nf.ca/articles/exploration/cooks-charts.php]; James Cook (1728–1779); Michael Lane (fl. 1763–1784).

8.   John Cabot (1450–*c*.1500). See Pope (2009).

9.   This is correct (John Stewart pers. comm.).

10. Serjeantson (2001). For the oldest fossils see Martin *et al.* (2000). Bourne (1993) made the same point about penguins being able to climb quite steep cliffs. For the soles of the great auk's feet, see Schulze-Hagen & Birkhead (2023) quoting Johann Friedrich Naumann (1780–1857) (1844).

11. Cartier (see Biggar 1924; Montevecchi & Tuck 1987): the bear may have been stranded via winter sea-ice, or it may have swum the 60km from the mainland.

12. Polar bears usually eat seals or walrus, and resort to seabirds only when they are desperate, as with Cartier's stranded bear, or when a lack of sea-ice makes seals inaccessible. This is happening with increasing frequency at the present time due to climate change.

13. Seebohm (1832–1895) quote: Seebohm (1885: 3: 372).

14. Henri Cosquer (b. 1950). Cosquer Cave (Cosquer *et al.* 1993). Note that there are several palaeo-images purported to be great auks – see Sigari *et al.* (2021).

15. Bahn & Pettitt (2016).

16. D'Errico (1994).

17. See Eastham & Eastham (1995).

18. Langeveld (2020). I queried this but Bram Langeveld convinced me by providing examples of cut marks on butchered bird bones. See also descriptions in Meldgaard (1988).
19. Armit (2006).
20. Tuck (1976).
21. In Montevecchi & Tuck (1987: 32). See also J. A. Tuck (1976) and the more recent website: ov.nl.ca/tcar/provincial-archaeology-ann ual-report-series/1997-toc/renour-bell-1997-pac-burial-site/
22. Avian gizzard stones: Ray (1768: 211), Ingersoll (1923: 95) and Birkhead *et al.* (2023).
23. William Cormack (1796–1868). See Pastore & Story (2003) for the story of Shawnawdithit.
24. Houlihan (1986); Montevecchi & Tuck (1987).
25. Montevecchi & Tuck (1987).
26. Montevecchi & Tuck (1987), and Matthew Holley (pers. comm.).
27. Greenland: see Meldgaard (1988). Note that there is a statement in Buffon (1793: 9, 334) that he incorrectly assumed to refer to great auks: 'the Greenlanders drive them upon the coast, and catch them with the hand, for these birds can neither run nor fly. They afford subsistence to the inhabitants during the months of February and March, at least at the mouth of the Ball River, for they do not resort to all the shores indiscriminately.' Great auks would not have been near the shoreline or a river mouth in these months, so this must refer to a different species.
28. Hans Poulsen Egede (1686–1758). Egede (1741).
29. Otto Fabricius (1744–1822) – see Meldgaard (1988).
30. Meldgaard (1988) points out the mistranslation in the original by Helms (1929), where it says 'windpipe'.
31. Fabricius, cited in Meldgaard (1988: 165, 177). The mass slaughter of great auks on Funk Island was for feathers for beds.
32. Meldgaard (1988). See also: Norrgrén (2020).
33. Meldgaard (1988: 161).
34. Fuller (1999: 163–166) gives a detailed account, mentioning that the Copenhagen bird was probably Naumann's winter bird.
35. This is Fuller's Egg 1, Hay Fenton's Egg; its origin is unknown (Fuller 1999: 245).
36. Whitbourne (1561–1634), Whitbourne (1622).

37. Berland (2020: 254).

38. 'Other writers' include: Lucas (1890), Cartier (1534), Hore (1536), Parkhurst (1578), Peckham (1583) and Whitbourne (1622), all cited in Gaskell (2000). Kearly's (1880) image of great auks walking up a narrow plank and throwing themselves into a boat helped to reinforce the idea of a docile, stupid bird complicit in its own destruction. Other stories of great auks being herded onto sails laid out on the ground are more plausible, as described by Richard Hore (https://content.wisconsinhistory.org/digital/collection/aj/id/2182), but I can find no image of that.

39. Hutchinson (2014); Montevecchi & Tuck (1987).

40. Aaron Thomas (1762–1799): Thomas, cited in Murray (1968) – note on his journal; see Glassford (2006) and https://journals.lib .unb.ca/index.php/nflds/article/view/5888/6899.   Subsequent great auk researchers Bill Montevecchi and Les Tuck considered Thomas's account to be distorted and exaggerated. Another knowledgeable seabird biologist, Bill Bourne (1993), however, on checking the veracity of some of Thomas's statements, together with details of the way southern hemisphere penguins were treated by mariners, was more inclined to believe him. John Milne (1850–1913).

41. Most seabirds are highly 'site tenacious', returning to breed where they have done so previously, rather than move somewhere else, unless their breeding site completely disappears, as Geirfuglasker did in 1830. Long-lived auks can afford a few years of breeding failure, so over the course of time tenacity pays off. One year when I was studying auks in Labrador, their normally predator-free islands were occupied by Arctic foxes. Unable to visit their breeding sites without fear of being killed, the birds simply sat in flocks on the sea just offshore, waiting, rather than flying off into the unknown to find somewhere else (Birkhead & Nettleship 1995). In just the same way, even after a 95% decline in numbers in the 1940s, Skomer's remaining guillemots stayed put.

42. Cartwright (1792, in Townsend 1911).

43. Bourne (1993) states – though on what basis I do not know – that on Fogo Island, some 50km (30 miles) from Funk Island and closer to the Newfoundland coast, great auks were still present around 1820.

44. https://www.wikiwand.com/en/Funk_Island: *Encyclopedia   of   Newfoundland and Labrador*. During his expedition Molloy found three mummified great auks. Peter Stuvitz (1806–1842); Thomas Molloy (no dates); Frederic Lucas (1852–1929).
45. Townsend (1930).
46. Lucas (1890: 497, 511).
47. Lucas (1890: 512).
48. Alfred Newton (1829–1907) later replaced the missing bones with those from another of his specimens (Owen 1865; Fuller 1999: 34). John Hancock (1808–1890); Richard Owen (1804–1892); Alexander Agassiz (1835–1910).
49. Lucas (1890: 493). Cat mummies as fertiliser (Martin 1981). Human bones for use in agriculture (Pollard 2021; Wilkin *et al.* 2023).

Chapter 3: The Auk and the Walrus

1. Keighley *et al.* (2019).
2. Nettleship & Evans (1985). They also list nine putative colonies.
3. Serjeantson (2001).
4. Warder Clyde Allee (1885–1955), see Allee & Bowen (1932).
5. Birkhead (1977, 2016).
6. Bardarson (1986) and Aevar Petersen (pers. comm.): two sites on the Miðnes peninsula, where Keflavik Airport is located, formerly called Rosmhvalanes, meaning Walrus Peninsula. See Keighley *et al.* (2019). The sites are *c.*8.5km (5 miles) apart; the harvesters for great auks came from this area.
7. Newton (1861), Sigurðsson (1770) in Bardason (1986).
8. Newton (1861: 382).
9. The 'treading down' quote is from Newton (1861), and that term had been used by many others visiting seabird colonies. Regarding the Foljambe Auk at Osberton Hall: this is named for George Savile Foljambe (1800–1869). Morris (2023) states: 'A note by Bartholomew Corbett in the Osberton papers says: 'The Great Auk skin was sent by Mr. Walker, a Liverpool merchant, as a present in October 1813 and was set up by me, being at Osberton at the time (in 1845).' From this, it appears that the bird was kept as a skin for 32 years before being set up as a full mount.

10. These are Frederik Faber (1796–1828); Count Frederik Christian Raben (1769–1838). See Newton (1861: 385) and Fuller (1999).
11. Newton (1861: 379).
12. Newton (1861: 389).
13. Ole Worm (1588–1654), Worm (1655); John Ray (1627–1705), Ray (1678); Francis Willughby (1635–1672).
14. Louis XIV (1638–1715), see de Lozoya *et al.* (2016).
15. Martin Martin (Màrtainn MacGille Mhàrtainn) (*c.*1660/69–1718). Fuller (1999); Bones (1993). As Andrew Fleming (2024) has said: 'By 1844, Satan would have been firmly installed in the [St Kildan] islanders' pantheon, as it were, of the forces of darkness. The emotional tenor of their experiences in church must have reinforced the St Kildans' belief in the power and ineffability of supernatural forces.'
16. Martin (1698).
17. Isabelle Charmantier, pers. comm.
18. See Schulze-Hagen & Birkhead (2023). After this paper went to press, I came across records for 1836 and 1838 from John Wolley's notebook (1: 125, transcribed by Gísli Pálsson). They are somewhat confused, as from an interview conducted on 17 May 1858, Wolley wrote:

> *Kjöbmand Targasen tells me that in 1838 three Geirfugl* [great auks – skins] *and two eggs were got at the Fuglesker, and that two years before viz. 1836 M. Guimard* [Joseph Paul Gaimard (1793–1858)] *of the French expedition got 5 skins in a booting* [meaning?]. *That was sank out by them ... Two or three days before this information young Knudson had told me* [*i.e.* John Wolley] *from Thorgersson that it was in 1836 that the 3 birds and 2 eggs were obtained by Chr. Hanson in Keflavik. It was in consequence of the solicitations of Hr. Knudson (the grandfather of my informant) that these skins and eggs were got from him. Hanson gave 80 Rd. for each bird. Knudson was residing at Keflavik from Copenhagen for the summer.*

Thus three, five or eight skins were obtained in 1836, or possibly three in 1836 and zero or three in 1838, or possibly just five in total. I checked Gaimard's nine-volume account of his journey to Iceland and Greenland (via the Biodiversity Heritage Library) and although he lists the great auk as occurring in Iceland, he makes no mention of any skins obtained. Similarly, there is no mention of great auk skins in Snaevaar's (2019) biography of Gaimard (Aevar Petersen, pers. comm.). Gaimard's diaries are held in the archives of the French military at Château de Vincennes outside Paris, and they may hold the answer. Judging from a map showing Gaimard's route, it does not appear that he stopped at Eldey, and he may therefore have obtained any great auks from Reykjavik or elsewhere. However, Alfred Newton, in a note added in pencil to Wolley's notes dated 1 July 1861 (*i.e.* after Wolley had died) said: 'I doubt if Gaimard ever had one' (skin, presumably). I'm grateful to Gísli Pálsson for this information.

Regarding 1840 or 1841: Newton says that 'the people of Kyrkjavogr did not visit Eldey in 1841 but that a certain Stephan Sveinsson did (the local people considered him a poacher) and he collected two birds and 1 egg that he later sold to a factor at Keflavik, Carl F. Thaae, who had three skins, a body in spirit and three eggs bought in 1841' (Newton's *Ootheca Wolleyana* part 2). This suggests that no birds were taken on Eldey in 1840.

19. Preyer (1841–1897). From William Preyer (1962).
20. From Newton (1861). For an account of the killing of the last two great auks, see Fuller (1999), Schulze-Hagen & Birkhead (2023) and Palsson (2024: 120). Carl Franz Siemsen (d. 1865).
21. Sven-Axel Bengtson (1944–2019); his statement (1984: 9) that the great auk was probably never numerous was rejected by Bill Bourne (1993), and I agree: the archaeological evidence shows just how abundant they once were.
22. Fuller (2001).
23. Timothy Fridtjof Flannery (b. 1956). Flannery (1994); see also McCaughley (2015).

Chapter 4: Three Men in a Boat

1. The naturalist François Juste Isidore Favier (1815–1867). As Newton said (*Ootheca Wolleyana* Vol 1: xiv), in Tangiers, Wolley 'unexpectedly found domiciled a keen egg-collector, at that time

known to few naturalists in Europe and perhaps to none in England'. Favier had worked as a taxidermist for the hunting parties of Edward William Auriol Drummond-Hay, Consul-General at Tangiers, Morocco.

2. Hugh Strickland (1811–1853). Strickland's monograph on the dodo was published in 1848. Five years later Strickland was killed after accidentally stepping into the path of an express train while trying to avoid a goods train on the other track.

3. Barclay Bevan had bought the egg from the ornithologist John Gould (1804–1881) (see Fuller 1999: 250).

4. *Ootheca Wolleyana* (Newton & Wolley 1860–1907).

5. John Wolley, cited in William Jardine's (1848–1853) *Contributions to Ornithology* 30: 4: 115.

6. Newton (1861).

7. The article, entitled 'The Great Auk still found in Iceland' by Lewis Lloyd, was in the *Edinburgh New Philosophical Journal* (1853–4, 56: 260–2). The same information is also reported in his *Scandinavian Adventures* (1854, ii: 495). Danish zoologist Japetus Steenstrup (1813-1897).

8. From Newton (1861). See also Palsson (2024). Geir T. Zoëga (1857–1928).

9. Palsson (2024: 116–17); Newton (1861) says '8 or 10 fish'. Gaskell (2000: 21) says '8 or 10 skildinga (shillings)' which makes rather more sense, but it isn't clear where 'shillings' came from. Newton's information came from Guðni Sigurðsson´s description of Geirfuglasker, thought to be written 1770–71. It was later transcribed by Jón Thorarensen (1929) who says in a footnote on page 88: 'I [*i.e.* Guðni Sigurðsson] have known Danish people paying 8 to 10 f(ishes) for one empty, blown egg.' So Thorarensen states that the abbreviation 'f' stands for fish, although this is not explained any further, and also implies that Newton and Wolley thought or had been told that 'f' stood for fish. In the Middle Ages, Icelandic people began to use the currency 'fish', instead of cloth, which they had used previously, in units of about 1kg. Fish was the most commonly used trading unit during the 18th century (Aevar Petersen, pers. comm.).

10. For images of eggs see Tomkinson & Tomkinson (1966) and Fuller (1999). J. W. Tomkinson (1912–1995) and P. M. L. Tomkinson (1919–2009). Just one of the 75 known eggs – Lord Garvagh's Footman's Egg – was emptied through a single drill hole on its side (Fuller 1999: 298 and M. T. Siva-Jothy, pers. comm.).

11. Newton (1861). See also Pálsson (2024: 170).

12. Newton (1861).

13. Tim Milsom, pers. comm.

14. George Dawson Rowley (1822–1878) was educated at Eton and at Trinity College, Cambridge. He is not well known among today's ornithologists; historians of ornithology, however, are aware of Rowley's idiosyncratic but wonderfully illustrated *Ornithological Miscellany* published between 1876 and 1878. As a person and an ornithologist, Rowley remains relatively obscure and underrated, and the numerous essays he wrote for his *Ornithological Miscellany* deserve better recognition and study.

15. Roberts (1861: 7353). Hewitson (1831–1838), Hewitson (1831), Champley (1829–1895), Champley (1864).

16. Newton (1861).

17. Victor Fatio (1838–1906), Fatio (1870). From my own research perspective, one of Fatio's most interesting ornithological discoveries was the extraordinarily large 'cloacal protuberance' of the male Alpine accentor, a structure containing the male's paired sperm stores (analogous to the paired epididymis in mammals), and a reflection of – as subsequent research revealed – the species' intensely promiscuous lifestyle.

18. Rowley (1995).

19. Newton (e.g. 1860, 1861, 1865). Notwithstanding Alfred Newton's pedantic perfectionism, he completely overlooked one of the most important great auk papers. Published in 1844, the year that Newton's bird became extinct, and more than a decade before his and Wolley's obsession began, J. F. Naumann's account of the life of the great auk would have reshaped Newton's vision of the bird. The omission was probably because Newton was only about 15 when Naumann's article was published. Also, communication between central Europe and Britain was so much poorer then, and Newton did not speak German (see Schulze-Hagen & Birkhead 2023).

20. Rowley (1995: 133).
21. Symington Grieve (1850–1932). Newton's (1885) review is harsh, without specifying exactly what errors Grieve is supposed to be guilty of. One might have thought Grieve would be devasted, but in fact a letter he wrote to a fellow great auk enthusiast, Robert Champley, on 31 August 1885 (having seen a proof of Newton's review, but before it was published), said that 'even though adverse, it has caused me such amusement as I am afraid the perfection desired and expected by the learned Professor (while it may be Newtonian) does not appertain to most mortals, of whom I am one.' Newton's pedantry was clearly well known. (Letter courtesy of Errol Fuller.)
22. Or possibly his son, Fydell (1851–1933). See Fuller (1999).
23. I'm surprised that Newton simply gave up on the idea of a great auk monograph. Why did he not say to himself, 'Well, I'll produce one that's better than Grieve's and without the mistakes'? It is perhaps a measure of Newton's personality that he gave up and diverted his energies into producing *Ootheca Wolleyana*, his memorial to John Wolley. Reverend F. C. R. Jourdain (1865–1940), F. G. Lupton (1881–1970), Harold Gowland (1899–1957).

Chapter 5: All Things from Eggs

1. Thomas (1983).
2. Tomkinson & Tomkinson (1966), Fuller (1999).
3. Fuller (1999: 244).
4. The 'renaissance' was in a particular type of research. See also *The Most Perfect Thing* (Birkhead 2016), which I wrote partly to counter the anti-egg, anti-science culture. The 2000 renaissance risks overlooking the huge volume of previous egg research, much of it motivated by commercial poultry production.
5. See Fuller (1999: 321–22). An image of this unusually coloured egg was published in 1888 after it was purchased by a St Malo shipowner, who bequeathed it to Comte Raoul de Baracé of Angers. A recent photograph of the egg, in the Museum of Comparative Zoology (MCZ) in Harvard, shows an egg whose green maculation is rather paler than the image in Fuller (1999), suggesting that it has probably

faded since it was originally painted. The other egg with green markings is William Barbour's Egg, also in the MCZ, which has much finer streaks of green maculation (Fuller 1999: 316; M. T. Siva-Jothy, pers. comm.).

6. Birkhead (2016).

7. Mary Anne Cust (née Boode) (1799–1882): https://en.wikipedia .org/wiki/Mary_Anne_Cust

8. The statement that her collection was superior to that in the British Museum comes from an article in the *Liverpool Mercury* of 30 May 1828: see: http://www.old-merseytimes.co.uk/leasowecastle.html

9. Wilhelm *et al.* (2001); see also Walsh *et al.* (2001).

10. Friesen *et al.* (1993); Taylor *et al.* (2011).

11. Beat Tschanz (1920–2013), Tschanz (1990).

12. Experiments that I conducted (Birkhead 1978) showed that razorbills do not recognise their own eggs. However, Paul Ingold, who studied both razorbills and common guillemots, told G. P. Baerends that they do (Ingold, pers. comm. to G. P. Baerends, cited in Baerends 1982), and cited as 'Ingold in prep.' in Tschanz (1990). However, I have been unable to locate any published information on this topic by Ingold. In addition, while common guillemots recognise both their egg (visually) and their chick (via vocalisations) even before hatching, it seems paradoxical that, if razorbills can discriminate their own egg, they do not – according to Ingold's (1973) research – recognise their chick until towards the end of the chick-rearing period.

13. Taverner (*c.*1680–1768), Taverner (1718).

14. Selfish gene theory predicts that it would not generally be in an individual's interest to incubate and rear any but its own egg and offspring.

15. These two eggs are Tuke's Egg, taken in 1840, and Newton's Egg, taken in 1841 (Fuller 1999: 317 and 253 respectively). The curious and very unusual spiral distortion in both eggs is visible in the black and white photographs taken by Bidwell in the 1890s, reproduced in Tomkinson & Tomkinson (1966). Tuke's Egg: plate 44 in Tomkinson & Tomkinson (1966). Newton's Egg: plate 32 in Tomkinson & Tomkinson (1966), and also in the painting by Henrik Grønvold (1858–1940) in Newton's *Ootheca Wolleyana* part 2 (1907),

reproduced in Fuller (1999: 253). It is worth noting that while female common guillemots typically lay eggs that are similar in colour, maculation and shape each time (Birkhead *et al.* 2017; 2021), razorbills do so only sometimes, or only some females (Lyngs 2020).

16. Arthur Cleveland Bent (1866–1954), Bent (1946).

17. D'Arcy Wentworth Thompson (1860–1948), Thompson (1917).

18. Personality types A & B. Yes, I know it is dated, but it works. See: https://en.wikipedia.org/wiki/Type_A_and_Type_B_personality_ theory#:~:text=Type%20A%20personality%20portrayed %20higher,novelty%20seeking%2C%20and%20harm %20avoidance.

19. Frank W. Preston (1896–1965). See Preston (1953), Mayfield (1989), Biggins *et al.* (2018).

20. Previous explanations for the guillemot's pyriform egg include (i) spinning like a top, and (ii) rolling in an arc, neither of which has stood up to scientific scrutiny. The greater stability of a pyriform egg is demonstrated in Birkhead *et al.* (2018).

21. Johannes Christopher Hagemann Reinhardt (1778–1845); for the letter, see Schulze-Hagen & Birkhead (2023).

22. Birkhead *et al.* (2020; 2021). We later discovered, from Thomas *et al.* (2017 Fig. 3b), that their assistant, Christina Barilaro, had taken skin samples from a lateral great auk brood patch. However, at the time, unaware of the existence of two brood patches, they simply thought that their great auk specimen had a bald patch (Christina Barilaro, pers. comm.).

23. John Henry Gurney (1819–1890), Gurney (1868). It is interesting that Bill Bourne (1993) was also sceptical about certain things Martin Martin wrote about the great auk.

24. Fuller (1999: 191) and Schulze-Hagen & Birkhead (2023).

25. J. A. Naumann (1795–1817). *Naturgeschichte der Land- und Wasservögel des nördlichen Deutschlands und der angränzenden Länder.* For details of Johann Friedrich Naumann (1780–1857) see Haffer (2007).

26. Heinrich Frank (1772–1843); Schulze-Hagen & Birkhead (2023).

27. This specimen is known now as Benicken's Auk. It was killed in the vicinity of Qeqertarsuatsiaat, Greenland, probably by a Greenlandic kayak hunter, in the winter of 1814/15 or 1815/16.

It was delivered to Mr Heilmann, director of the Royal Greenlandic Trading Company in Qeqertarsuatsiaat. Five years later the skin turned up in Schleswig (then Slesvig, southern Denmark) in the possession of the ornithologist Johann Casimir Benicken (1782–1838). He had probably received it through one of the trading company directors. Some time between 1830 and 1842, the ornithologist Emil Hage, from Stege, Denmark, purchased the skin from Benicken, and on 20 December 1842 Hage sold it to J. Reinhardt, the director of the Zoological Museum in Copenhagen.

28. Naumann (1844).
29. This is from MacLeod to Mackenzie letter in 1776–77, cited in Bones (1993).
30. The Icelandic farmer's observation is reported in Preyer (1862).

Chapter 6: The Chick That Never Was

1. The term 'fledge' – first used in the 1300s and with the same root as 'fletch', referring to an arrow – applies to a young bird that is 'fit to fly'. No chicks of the various auk species can fly when they leave the breeding site because their wings are not fully developed, so strictly speaking the term 'fledge' is not appropriate. I use it here simply for convenience.
2. The three fledging strategies are described in Gaston & Jones (1998). Disliking the established scheme, Gaskell (2004) presented a revised version referring to intermediate species as 'quasi-precocial'. He categorised the great auk chick as precocial, but as explained in the text here, this was based largely on John Gould's erroneous account. It is not known why only the father accompanies the chick in these species. The female partner often returns to the breeding site for up to two weeks, possibly to defend it from other individuals looking for a place to breed in future years.
3. Martin (1698).
4. Birkhead (1993).
5. See Smith's (2006) account of the interaction following Gould's misrepresentation of Jemima Blackburn's (1823–1909) observation of a recently hatched cuckoo ejecting a meadow pipit chick. Further, unable to accept Darwin's ideas about natural and sexual

selection, Gould – it has been asserted – misleadingly portrayed many birds known to be polygamous as harmonious, monogamous family units (Birkhead 2022a).

6. Gould (1837: *Birds of Europe* 5), and Gaskell (2004). Elizabeth Gould (1804–1841).

7. Gould (1837: *Birds of Europe* 5).

8. In the 1860s, it was widely believed that the great auk was an Arctic species, even though no one had actually reported seeing them sitting on ice floes. It isn't (see Evans & Nettleship 1985). Even in the 1830s there was a substantial literature on the great auk on which Gould could have based a more accurate account. Martin (1698).

9. Gaskell (2000: 153).

10. My information on Raben's diary is from Aevar Petersen, pers. comm., and from Pálsson (2024).

11. This is from Audubon's *The Eggers of Labrador* (Audubon & Coues 1897). Destroying all existing eggs in order to obtain fresh ones over the next few days was routine. Aaron Thomas, writing in 1794, said: 'If you go to the Funks for eggs to be certain of getting them fresh you pursue the following Rule: You drive, knock and Shove the poor Penguins [great auks] in Heaps! You then scrape all the Eggs in Tumps, in the same manner you would a Heap of Apples in an Orchard in Herefordshire. Numbers of these Eggs, from being dropped [laid] some time, are stale and useless, but you having cleared a space of ground the circumference of which is equal to the quantity of Eggs you want, you retire for a day or two behind some Rock at the end of which time you will find plenty of Eggs – fresh for certain! – on the place where you had before cleared.' (Cited in Murray 1968). See https://www.gutenberg.org/files /39979/39979-h/39979-h.htm#Page_406.

Note that following certain previous authors such as Grenfell (1909), Gaskell (2000: 165) assumes these are black guillemots. This is incorrect; these were common guillemots. In contrast, black guillemots breed in crevices and under rocks so it would be very difficult to trample their eggs in the way Audubon describes. Pope (2009) states that fishermen used great auks and other seabirds to bait their traps in pursuit of cod, prior to the appearance of the capelin. Their use as bait was yet another pressure on Newfoundland's seabirds from the 1500s.

12. Newton (1861: 385), my italics.

13. Birkhead (1993).

14. The great auk probably weighed 3,560g (7.85lb) and its egg 351g (12.4oz), about 10% of the adult body mass and similar to its guillemot and razorbill cousins (Montgomerie & Birkhead, in press.).

15. Birkhead (1993), Gaston & Jones (1998) and Gaskell (2000, 2004).

16. Houston *et al.* (2010).

17. But as the great auk's distinctive pattern of wing moult shows (see p. 32), a phylogenetic constraint can be broken. The great auk's closest relatives are the razorbill, common guillemot and Brünnich's guillemot (see Smith & Clarke 2015).

18. Lucas (1890: 512).

19. Hobson & Montevecchi (1991).

20. Chris Milensky, pers. comm.

Chapter 7: Playboy, Pilot and Ornithologist

1. From Hewitt's (1923) account in the *Oologists' Record*, and Hywel's (1973) biography of Hewitt (120, 319–324). The biography was written at the request of Jack Parry. William Hywel Jones (1909–2001) used the *nom de plume* 'William Hywel'. Here, I cite this biography as Hywel, but when referring to the man, I use William Hywel Jones or just Hywel Jones.

2. Cole & Trobe (2000).

3. Fuller (1999: 266); see also Sotheby's site: https://www.sothebys .com/en/buy/auction/2021/natural-history-2/an-egg-of-the -extinct-great-auk

4. Hywel (1973: 48). Thomas William Good Hewitt (1868–1930) was Vivian's uncle. The Hewitt Brothers Ltd brewery was established in 1885.

5. Hywel (1973: 48).

6. Hywel (1973: 73).

7. Egg-collecting: see Cole & Trobe (2000), Cole (2015) and Hywel (1973: 127). The start of Hewitt's egg-collecting is documented in a set of notes headed '1920 – Bird Notes – 1921' in what have been referred to as the Hewitt Papers (Birkhead, Clugston & Fuller

2023), where he states that he collected guillemot eggs from Worms Head in south Wales in 1920, and at Puffin Island in 1921. When Mrs Parry and her children went to stay at Penmon with Hewitt, Mrs Parry's husband, John, did not accompany them. As well as the image in the text, there's another of Hewitt and her – with a long plait of fair hair reaching far down her back – standing together on a cliff top.

8. Charles Harold Gowland (1899–1957). Between that first letter in 1924 and 20 September 1955, Gowland wrote no fewer than 351 letters or telegrams to Vivian Hewitt. Few of Hewitt's replies exist. For most of the correspondence, Hewitt and Gowland address each other formally: 'Dear Captain Hewitt' and 'My dear Gowland'. Then in 1946, Gowland addresses Hewitt as 'Dear Skipper' (28 October), or 'My dear Skipper' (30 November) and signs himself 'Harold' rather than C. H. Gowland as he had done previously.

9. Hewitt Papers.

10. Whitaker (2021); some of these eggs are now in the Alfred Denny (Zoology) Museum at the University of Sheffield.

11. Verity Peterson, F. G. Lupton's granddaughter, pers. comm.

12. Birkhead (2022b).

13. D. Cotgrave, pers. comm.; see also Cole & Trobe (2000) and Lovegrove (1990) for the kite's tale.

14. Jack Parry (1909–1998), Myfanwy (Girlie) Parry (1912–?), Paul Kenneth (Ken) Parry (1911–?) and Vivian Weston Parry (1914–1942), their mother, Nellie (Eleanor Myfanwy, née Jones) Parry (1883–1969).

15. Hewitt Papers.

16. Hywel (1973).

17. Public Records.

18. Chilton (2009: 296).

19. As far as I can tell, Hewitt did not buy Gowland's great auk egg, but the seed was sown.

Chapter 8: Spend, Spend, Spend

1. Parkin (1911: 7–8).

2. Rowley's six eggs and who bought them. The two damaged eggs were the Bowman Labrey Egg mentioned in the prologue, and one

previously owned by Lord Garvagh. The Reverend Francis Jourdain bought the damaged and (badly) repaired Bowman Labrey Egg (140 guineas), and another intact specimen known as Lady Cust's Egg (210 guineas). Of the two remaining eggs, Captain Cook's Egg went to a Mr Bernard Eckstein for 260 guineas, and the other, the broken and repaired Lord Garvagh's Footman's Egg, went to a Mr G. N. Carter for 100 guineas (Jourdain 1934).

3.   It was Errol Fuller who in the 1990s identified the auk specimens that could have been the last two.

4.   Note that this is not the male sold by Israel! The price of great auk relics over time (up to 1970) has been documented by Bourne (1993). Both eggs and skins rose sharply in price after the 1850s, with the price of eggs peaking in the early 1900s while skins were still increasing in 1970. Prices of both have remained high ever since.

5.   Hewitt Papers.

6.   Hywel (1973: 159).

7.   Hywel (1973: 159).

8.   Hewitt Papers.

9.   Jourdain (1934).

10. Hewitt Papers. On 16 September 2020, I was invited to meet David Clugston (DC) and Steve Holloway (SH) at David Wilson's house in Aylesbury to collect the guillemot eggs that DC and SH had donated to the Alfred Denny Museum. I returned to Wilson's house on my own (with their permission) on 17 March 2021 to retrieve some additional eggs they had donated. It was while I was in the poorly lit, cramped and cluttered attic that I found an egg cabinet previously overlooked by the others. All the drawers from the top were empty, but then I came across a tray of 12 great auk eggs. My heart literally missed a beat. Was this a haul of previously unknown great auk eggs? My excitement evaporated on picking one up. I knew instantly from its weight that it was a replica, as were the others. They were very good. Until July 2023, DC and I were uncertain where these eggs came from. We knew that they were not the same replica eggs that Hewitt bought from F. G. Lupton in 1935. A note in the tray, however, said 'from Goodall'. The mystery was solved when DC found two letters in the Hewitt Papers: one

from Jeremiah Goodall (1862–1939) and the other from his brother, Steve (1860–1930). It was Steve Goodall – an inspector for Trinity House – who, having found the wooden bowsprit from a ship, realised that this wood, yellow pine, was the perfect material from which to make the 12 replica eggs. The eggs passed from Steve Goodall to his brother Jeremiah, whose collection Vivian Hewitt purchased in 1935, and whose collection in turn was rescued after his death by David Wilson, from where the replica eggs (four each) passed to DC, SH and myself in 2021. After Steve had made the 12 replica eggs in 1922, Jeremiah Goodall got him to make a further seven replicas. These seven were sold via Stevens's Auction Rooms to F. G. Lupton for £15 in 1925, and then sold on to Vivian Hewitt for £20 in 1935 (Hewitt Papers). Their whereabouts remain unknown (see Birkhead *et al.* 2024).

11. Which Seebohm (1891) calls the Museum of the Royal College of Surgeons, saying that this egg had not previously been illustrated. There are in fact two great auk eggs in Seebohm (1891), both from the Royal College of Surgeons; the other is Jack Gibson's Egg. Seebohm does not say who the artist is, and Tim Milsom, Seebohm's biographer, did not know (pers. comm.).

12. John Hunter (1728–1793). Newton (1865) pursued Dr Dick, but without success. His only clue was that someone named 'Dick' had bought a 'parcel of eggs' on 11 June 1806 during the sale of Sir Ashton Lever's (1729–1788) Museum known as the Holophusikon (a reference to its universal coverage of natural history). The Leverian Museum, as it was more conveniently called, was famous for its biological and ethnographic material, including some items obtained by Captain James Cook during his voyages.

13. Bourne (1993): https://www.pdavis.nl/Surgeons_1840.htm. Bourne suggests another possibility, which is that the eggs may have been obtained by John Hunter himself for his museum when he visited Cape Breton and was staff surgeon on expedition to the French island of Belle Île during the Seven Years War with France. But Bourne seems to have got his chalazae in a twist here, for as far as we know, Hunter never visited Cape Breton (Wendy Moore, author of *The Knife Man,* a biography of John Hunter, pers. comm.).

14. Cartwright's journal (1792). See Townsend (1911).

15. Cartwright's Townsend journal (1911: 23–24). See also Thrush (2016). In 1830, Captain Robert Fitzroy (1805–1865) brought four Fuegians back to Britain from his *Beagle* voyage with Darwin.

16. In 1766, Joseph Banks (aged just 23) travelled aboard HMS *Niger* to Newfoundland and Labrador to study the country's natural history. He reported seeing large numbers of 'penguins' (great auks) around the boat on the Grand Banks, and in Chateau Bay, Labrador, he obtained 'a specimen'.

17. The four Inuit that died of smallpox were cremated and their ashes buried in Plymouth. Cartwright's account of this (in Townsend 1911; see also Thrush 2016) is heart-rending.

18. Dr Elisha Cullen Dick (1762–1825); see Moore (2006: 122). The two other eggs were Jack Gibson's Egg and Alfred Newton's Egg (Fuller 1999).

19. Hewitt Papers.

20. Hewitt Papers. When I visited Bryn Aber in 2023, some of these sheds were still extant and still contained some of Hewitt's belongings. All the eggs and skins had gone, but there were empty cabinets and masses of machine and car parts.

21. These are the Cincinnati Egg and Wallace Hewett's Egg (Fuller 1999).

22. Hywel (1973: 189); he lists (p. 155) the following individuals whose collections Hewitt acquired. I have added first names, birth and death dates and profession, the latter to illustrate the wide range of respectable occupations of egg-collectors prior to collecting becoming illegal in the UK in 1954. They were all part of a network of oologists.

    • Bennett, Arthur George (1888–1954): customs officer, Falklands
    • Goodall, Jeremiah Matthew (1862–1939): mining engineer, South America (Cole & Trobe 2000: 97)
    • Jourdain, Rev. Francis Charles Robert (1865–1940): clergyman (Cole & Trobe 2000: 125)
    • Massey, Herbert (1852–1939): textile business (Cole & Trobe 2000: 167)

- Pybus, William Mark (1851–1954): lawyer, Newcastle upon Tyne (*Nature* 1924: 113: 169)
- Stuart-Baker, Edward Charles (1840–1944): police force, India (Cole & Trobe 2000: 247)
- Swann, Harry Kirke (1871–1926): book dealer (Cole & Trobe 2000: 249)
- Wilkes, Arthur Hamilton Paget (1898–1955): clergyman/ missionary in Africa
- Lupton, Frederick George (1881–1970): solicitor, Accrington, Lancashire

23. Nina Bagley: https://mybeautfulthings.com/2021/07/14/a-poem -a-poppy-and-a-new-outfit/gathering-by-nina-bagley/
24. 'Satisfaction renews' is from Harrison's (1996) last book, written as she was dying of cancer, and said to be little more than a stream of consciousness, but an intriguing one nonetheless. Gabriel Garcia Marquez (1927–2014); Barbara Grizzuti Harrison (1934–2002).
25. The collector rules: quote is from Emma Smith (2022: 120); Billie is referred to in Hywel (1973: 23–26).
26. Ornithologist and egg-collector Alexander Koenig (1858–1940), founder of Berlin's Koenig Museum in 1912, suffered in his advanced age (in the 1930s) from tremors (perhaps due to Parkinson's), which ceased when he held an egg in his hand. 'My early mentor Heinz Mildenberger, a famous nest finder and oologist (like his father) had good connections to Koenig because of his bird egg interests. Heinz M. was deeply impressed by this and told me the story repeatedly' (Karl Schulze-Hagen, pers. comm.).
27. Paul Singer (1904–1997); Keefe (2021: 69, 70, 75, quoted with permission).
28. Hywel (1973: 168).
29. Hywel (1973: 191, 199, 200, 201).
30. The Hewitt Papers include his will. Interestingly, Hewitt and Jourdain wondered about employing Desmond Nethersole-Thompson (1908–1989) to look after their putative museum. Like Jourdain, Nethersole-Thompson was a respected ornithologist/ egg-collector.

31. Hywel (1973) and Fuller (1999: 392). Pitman collected eggs in Africa and Palestine (Cole & Trobe 2000). In the 1940s, William Glegg (1878–1952) was an 'unofficial worker' – an honorary curator – at the Tring Museum, cataloguing Rothschild's vast collection of eggs. On 18 August 1945, Glegg wrote to Hewitt to tell him first that he knew, through Francis Jourdain, that Hewitt was planning to create an oological institute. Second, he said that the Museum had recently decided to appoint a curator of birds' eggs, and third, that Glegg had suggested to his boss, Norman Kinnear (1882–1957), that at least part of the Tring Museum 'should be used as an oological institute'. Glegg's letter to Hewitt was to tell him that 'there may exist now an opening for oology which may never occur again'. Of course, it came to nothing in terms of Hewitt's entire egg collection, although some of Hewitt's eggs are now housed there (Hewitt Papers).

32. John Rocke (1817–1881); Rowland Ward (1848–1912). Fuller (1999: 126).

33. The Malcolm family had made its fortune in Jamaica from sugar and slaves. John Malcolm (1805–1893), the 14th feudal baron of Poltalloch Argyll (see: https://mssprovenance.blogspot.com/2020/08/the-illuminations-of-john-malcolm.html), became 'one of the leading collectors of old masters of his day'. A couple of years before heading off to the Highlands for Hewitt, Peter Adolph (1916–1994) invented the table football game of Subbuteo that later made him wealthy. He subsequently travelled widely in Europe and the USSR collecting the eggs of various species, but especially warblers. After a colourful and inventive personal life, he died aged 77 after an accident [https://thebeautifulworldoffootball.wordpress.com/2022/04/09/growing-up-with-subbuteo-mark-adolph/]. The great auk egg bought by Hewitt was named Malcolm's Egg by Fuller (1999).

34. Benjamin Leadbeater (1771–1851). Dodo – see Hume (2006).

35. Bond (1903); Fuller (1999).

36. Edward Lear's great auk is in Gould's *Birds of Europe* (1832–1837). Jizz – see Greenwood & Greenwood (2018). Lear's (1812–1888) image of himself as a great auk appears in a letter to Lord Carlingford in 1863.

37. J. C. Squire, cited in Moss (2013: 139). J. C. Squire (1884–1958) was a writer and editor of the literary magazine *The London Mercury*. His statement about the change in the public attitude towards egg-collecting appears in his introduction to E. W. Hendy's book *The Lure of Bird Watching* (1928).

38. The identity of Tucker is unknown; it is unlikely to be the ornithologist Bernard Tucker (1901–1950) (Ian Newton, pers. comm.). Gowland's Tucker may have been a member of the 'Association of Birdwatchers and Wardens' established by Nat Tracey, who lived in the Kings Lynn area at the same time as Tucker (Sean Nixon, pers. comm.). Gowland letter in the Hewitt Papers.

39. Hewitt Papers.

40. Correspondence between Gowland and Hewitt, May 1937, in the Hewitt Papers.

41. Hewitt Papers.

42. Hewitt paid a total of £2,200 (about £102,000/$130,000 today) for Jourdain's collection (£1,500 for the eggs and library comprising 888 volumes, and £700 for the two great auk eggs) – an incredible bargain. The acquisition was messy and, as Hewitt said, unsatisfactory. Before purchasing it, Hewitt employed the retired customs officer and naturalist Arthur Bennett (1888–1954) to assess the collection for him. Bennett did a great job, sending Hewitt two long hand-written letters on 16 and 21 June 1940, documenting all of Jourdain's eggs and books respectively. In the second letter, Bennett told Hewitt: 'Alexander (of *Ocean Birds*) of Oxford University was here yesterday and I find he had already taken all manuscript books diaries & certain foreign Bird Books and all photos and negatives.' Wilfred Backhouse [W. B.] Alexander (1885–1965) was based at the Edward Grey Institute, Oxford, and his personal collection of bird books provided the nucleus of the Alexander Library, which would become one of the best ornithological libraries in the world (see Cheke 2020). W. B. Alexander became its librarian in 1945. Hewitt was incensed by the fact that Alexander had taken the Jourdain material, and initially refused to purchase the rest until the dairies were returned. They weren't, but Hewitt eventually bought the collection anyway.

'Surified' probably means pilfered or broken (according to Gowland's letter of 15 December 1947). Jourdain's daughter, who did not approve of his collecting, was in charge of his material after his death. She did not take care of it; some of the trays of eggs were damaged, and she sold off or gave away some of Jourdain's books and diaries (see Mearns & Mearns 1998). It must have been his daughter who allowed W. B. Alexander to take Jourdain's notebooks.

43. Letter dated 21 January 1946 in the Hewitt Papers.
44. Hewitt Papers.
45. Hywel (1973: 277).
46. Hywel (1973) and Hewitt Papers. In 1947, Girlie and her husband moved into Tŷ'n Llan farm, which Hewitt had bought previously but leased to tenants (Hywel 1973: 142, 296).
47. Hywel (1973: 295).
48. Nixon (2022: 51). Well-known and much-respected naturalists, including Sir David Attenborough and Mark Cocker, had collected eggs when they were boys and acknowledge the role it played in developing their natural history and conservation interests.
49. Grigson (1956).
50. Nixon (2022: 58).
51. Information about Frank and Olive's affair was told to Jim Whitaker by the late J. A. Yelland, and Yelland told me (pers. comm.) the same on 1 October 2022, shortly before he died. The RSPB information is from Anonymous (1955, 1956). Gowland died aged 57, during a cricket match while playing with his son (Cole 2006).
52. Given that Hewitt's passion for birds' eggs was so well known, it is surprising that he wasn't 'on the list', as Gowland warned. But the authorities may have been wary: if you are going to pick a fight with someone, you do so only when you think you are going to win – just as in the Sackler story.

Chapter 9: Post-mortem Fate

1. Hewitt Papers and Hywel (1973). Ken Parry had lost his wife, Barbara (the date of her death is not known), and went to live with Hewitt (Hywel 1973: 303). Not informing a terminally ill patient about their condition seems very odd, but the family correspondence

is clear. I asked a consultant surgeon friend, and he said that in the 1960s it was routine not to inform terminally ill cancer patients.

2. Hewitt Papers and Hywel (1973). 'Mother is quite lost'; letter from Ken to Jack, 24 June 1965. Pat Venables (1903–1989).

3. Hywel (1973: 298).

4. Hywel (1973: 302, 307).

5. Hewitt Papers.

6. David Wilson (1926–2020). Most of Jourdain's notebooks were acquired by W. B. Alexander soon after Hewitt's death, before Hewitt could decide whether to purchase Jourdain's collection. Hewitt fought hard for their return, without success. See Chapter 8, note 43. Hewitt Papers.

7. The BTO solicitors were Vaisey & Turner.

8. Dick C. Homes (1913–1978). Age is relative; in October 1965 Wilson was 39 and Jack Parry was 56.

9. Hewitt Papers. Reginald Wagstaffe (1907–1983); Charles Tunnicliffe (1901–1979).

10. James Macdonald (1908–2002). 'At present printing', this is Tomkinson & Tomkinson (1966). Kreuger, Helsinki: Ragnar Kreuger (1897–1997) was a Finnish industrialist and egg-collector. In 1962, he donated his collection to the University of Helsinki's Finnish Museum of Natural History as 'Museum Oologicum R. Kreuger'. See note 26 from Chapter 8 above.

11. Hewitt Papers: Mrs Parry died of a stroke on 29 September 1969, in the Bahamas.

12. Walter Rothschild (1868–1937). Colin J. O. Harrison (1926–2003); Kenneth Williamson (1914–1977). An embarrassment of riches. Imagine the outcry if, on being offered the contents of Tutankhamen's tomb, the British Museum had said, 'we have no space for all of this, take some and give the rest away to anyone who wants it'.

13. Jeremy Greenwood, pers. comm. John Eleuthère du Pont (1938–2010).

14. The Alfred Denny Museum: https://www.sheffield.ac.uk/alfred -denny-museum.

15. Anonymous (1968). See also Birkhead *et al.* (2023).

16. Clugston (2021).

17. See Morris (2003) for information on fake great auks.
18. https://antique-collecting.co.uk/2020/10/21/great-auk-flies-in
     -gloucestershire-sale/

Chapter 10: Witch Hunters

1. Hywel (1973: 172). Jim Flegg (b. 1937).
2. To their huge credit, curators like Bob McGowan (b. 1955)
    (National Museum of Scotland) and Douglas Russell (b. 1972)
    (Natural History Museum at Tring), and others, have been proactive
    in preserving egg collections they have heard about or have been
    donated, rather than allowing them to be ground to dust as some
    conservation bodies advocate. In this way, these collections
    continue to provide a valuable resource for researchers (see
    McGowan 2021).
3. Elephant birds existed on Madagascar until 1000–1200 AD. Richard
    Ford once worked for Watkins and Doncaster, the Natural History
    dealers, but we seem to know little else about him. There's a story
    to the Russian ostrich; this particular egg is a plaster cast of a fossil
    egg of the extinct ostrich *Struthiolithus chersonensis*, found in the
    Gobi Desert. The cast was given to a Russian museum from which
    it was looted during revolutionary times. It then found its way to
    northern China, where it was bought as a souvenir by George
    Mason and brought to England, and eventually ended up in Hewitt's
    collection (Ford 1968). As regards the sale at Spink & Son, this
    began as an auction, and anything not sold immediately could be
    bought later. Hence the four years it took to sell all the eggs.
4. Fuller (1999).
5. As a young man in 1894, Wallace Hewett (about whom little is
    known) bought two great auk eggs at an auction in Rochester,
    Kent. He was the only bidder, and, it seems, the only person to
    recognise the eggs in a box of curiosities. After purchasing the box
    and its two great auk eggs for 36 shillings, he quickly sold them –
    one for 175 guineas, the other for 260 guineas (Fuller 1999: 324).
6. Fuller (1999) and pers. comm.
7. Clugston (2020).
8. Jack Gibson (1926–2013). Gibson (1984).
9. Hewitt Papers: letter from Gowland to J. Gibson.

10. Bob McGowan, pers. comm. Other evidence that Gibson was, like so many of his era, an egg-collector, comes from a recent study of the history of the extremely rare Slavonian (or 'horned' in the US and elsewhere) grebe in Britain: Gibson took 10 clutches in 1961 – that is, several years after the law had changed (Benn *et al.* 2023). Ian D. Pennie (1916–2002); Arthur Clarke (*c.*1923–2012).

11. Clugston (2013).

12. This is based on a 420-word note Wilson wrote for himself describing his visit to Gibson on 12 November 1993: 'Great Auk Eggs, 8 seen at JAG's on 12.11.93'. Gibson stated that the 'wife of a director' that owned the eggs was none other than David Spink's wife Dorothy (1916–2006), and that she had renounced alcohol, taken up religion and 'is to found a religious institution in France'. The cash was to comprise '£4,000 (duly receipted to Spink) and £10,000 in used bank notes (duly delivered to Spink)'. Gibson asked the Marquis of Bute if he would 'stump up £10,000, if JAG produced the £4,000', but Bute died, so all the cash came from Gibson, but in the name of the Scottish Natural History Library – or at least this would be the official position as he 'could not possibly risk being called an egg-collector'. Hewitt Papers.

13. At that time the science research councils would not have funded a project on the history of ornithology, but they would, and did, fund my research on avian mating systems. I later discovered that the Leverhulme Trust had a broader outlook, and they subsequently funded much of my research that embraced both the history and science of ornithology.

14. Hewitt Papers: letter to Armand Hammer (1898–1990) dated 18 January 1986: £350,000.

15. Hewitt Papers: David Wilson's note to himself ('12.11.93') after visiting Jack Gibson and seeing the eight eggs, and a carbon copy of the letter Wilson sent to Jack Gibson on 18 November 1993.

16. Fuller (1999) and pers. comm. Errol Fuller told me in November 2022 how his own interest in the great auk started around the age of eight when his grandmother showed him a book – probably Wood's (1882) *Illustrated Natural History of Birds* – with an account of the great auk's extinction.

17. For example, Fuller (1999: 274). When they met on 12 November 1993 (note 12 above), Gibson told David Wilson that he would give the eggs to the Scottish National Museum.

18. Sheikh Saud Al Thani (1966–2014) was the founder of Al Wabra Nature Preservation, which at one time held most of the world's Spix's macaws. My colleague Nicola Hemmings and I helped Al Wabra assess the causes of hatching failure in their (highly inbred) Spix's macaws in 2010 (Hemmings *et al.* 2012).

19. Information from Errol Fuller (pers. comm.) and also from Adam & Kerr (2014). After Sheikh Saud had paid Fuller for the one egg, Fuller was left with a debt of around £50,000 (Errol Fuller, pers. comm.).

20. The reserve bid was £40,000–£60,000 but it went for £101,000 to someone in the Far East. The Natural History Museum at Tring also bid for the egg but was unsuccessful (Errol Fuller, pers. comm.). See https://www.sothebys.com/en/buy/auction/2023/emma-ha wkins-a-natural-world/lady-custs-great-auk-egg.

21. https://www.sothebys.com/en/buy/auction/2021/natural-history -2/an-egg-of-the-extinct-great-auk. The information that it was not sold was from Errol Fuller (pers. comm.).

22. Birkhead *et al.* (2023). There may be another dimension to this, since in late December 2012, six months before he died, Gibson prepared a statement announcing his winding-up of the library. In it he states that 'the Scottish Natural History Library was never intended to be a permanent institution. Indeed, as can be seen from the Constitution and the Annual Summary of Activities, the ultimate aim was, as soon as the Council considered the Library to be as complete as possible, to present all the significant holdings belonging to the Library as gifts in whole or in part, to fill gaps in the holdings of other Scottish educational institutions.' He then states that these aims were destroyed by the 2008 financial crash and digitisation. In an attempt to offset the consequences of the crash, Gibson said that he used £25,000 of his own money, albeit to no effect. Second, the digitisation of academic journals, which was well under way in 2012, meant that libraries no longer sought long runs of hard-copy journals. By late December 2012, Gibson's library had just £21.45 remaining, a sum he suggested donating to the Red Cross. Unpublished document in the Hewitt Collection.

Chapter 11: Afterlife Lessons

1. Information from https://en.wikipedia.org/wiki/Gannet_Islands _Ecological_Reserve, which states that the islands were named after a 19th-century ship, but this is incorrect, since the name Gannet Islands appears on a map made by Captain Cook's colleague Michael Lane in 1771 (reproduced in Townsend 1911). The origin of the name remains a mystery (see Birkhead 1993).

2. Jordan & Olson (1982).

3. A captive breeding programme. Even if there were birds to bring into captivity, it is hard to imagine great auks breeding sufficiently successfully to fuel a reintroduction programme. Guillemots have laid eggs in captivity but only rarely reared chicks successfully. Another critically endangered bird, the California condor, *has* bred successfully in captivity, with offspring released back into the wild (Walters *et al.* 2010). See also McCarthy (2021).

4. In September 1845, two birds thought to be great auks were seen in Belfast Bay, Ireland. Cork (Fuller 1999: 406); Gaskell (2000: 149); Skye (Fuller 1999: 406–07); Belfast Bay (Gaskell 2000: 143); Vardo (Fuller 1999: 408).

5. Newton (1861); Gaskell (2000: 146). Colonel Henry Maurice Drummond-Hay (1814–1896).

6. This assumes that the great auk was similar in the timing of its moult to the razorbill: see Harris & Wanless (1990). See also J. F. Naumann's painting of Benicken's auk on p. 109.

7. In 1872 an Icelandic seaman wrote to Alfred Newton to say that on 27 June 1869 he had seen two great auks 'near the island of Skrúdur off Eskifjördur [East Iceland], which due to storm and rough seas I could not capture' (Pálsson 2024: 210).

8. In 1936, nine king penguins arrived in Norway on the SS *Neptune*. The birds had been provided by Consul Lars Christensen (1884–1965), who was involved in whaling in Antarctica. The National Association for Natural Conservation had organised the whole scheme. Two pairs of penguins were released at Røst in Lofoten, and two pairs and one juvenile were released at 'Gjesvær in Finnmark … last seen in 1949'. The Antarctic was seen as a 'faunal resource for Scandinavia' and Lars Christensen's aim was to establish

the penguins so they could be 'farmed' for their (less than palatable) meat. They were actually last seen in 1948 (Rob Barrett, pers. comm.).

9. Fitzpatrick *et al.* (2005).

10. Newton (1896: 182).

11. Nicholson (1926: 242, 260). Edward Max Nicholson (1904–2003).

12. Nicholson (1926: 25). One might have hoped that egg-collecting was a thing of the past, but, as Joshua Hammer (2020) has engagingly and convincingly demonstrated in *The Falcon Thief*, in the decades following Vivian Hewitt's death, egg-collectors continued to take the eggs of rare birds, both for their shells and also for their embryos – that is, the chicks that would emerge from them. In a world where wildlife faces so many anthropogenic threats, the scale of wildlife crime is hugely depressing.

13. https://www.nhm.ac.uk/discover/news/2018/march/a-journey -through-the-largest-egg-collection-in-the-world.html

14. Martin *et al.* (2000). There are other long-extinct flightless auks, including one named after great auk enthusiast Alfred Newton, *Pinguinis alfrednewtoni* – a close relative of the great auk, as its generic name indicates. In a twist of fate, Jim Martin (1972–) became a publisher at Bloomsbury Publishing, where he commissioned me to write this book, 24 years on.

15. Jessica Thomas and John Stewart (pers. comm.), and Thomas *et al.* (2018). In contrast to the absence of any loss of genetic diversity in great auks, a study of whales using the same approach did show just this effect. Armed with information about the great auk's genome, there is another opportunity, in principle at least, to check whether, hidden in some remote corners of the North Atlantic, great auks still exist. Our molecular abilities are now so sophisticated we can take a sample of soil or water and simply scan it for the presence of DNA from particular species. Tiny traces of environmental DNA (eDNA as it is called) from skin, faeces, mucus, eggs, feathers, etc., allow the presence of a species to be detected. This new technology opens a new way of measuring biodiversity. In the great auk's case, proving a negative could be an issue.

16. Thomas *et al.* (2017).

17. Jessica Thomas, pers. comm.

18. Anthropogenic threats to seabirds, see Camphuysen (2022), Cunningham *et al.* (2022) and Paleczny *et al.* (2015).
19. See Frederickson (2010); Fayet *et al.* (2021).
20. Anonymous (2013).

Epilogue: A Less Perishable Inheritance
1. Pálsson (2024) and Gísli Pálsson, pers. comm.
2. https://www.iucnredlist.org/species/22704831/209015125 #threats
3. http://datazone.birdlife.org/species/factsheet/22685245

Bibliography

Adam, G. & Kerr, S. 2014. Saud bin Mohammed bin Ali al-Thani, collector, 1966–2014. *Financial Times*, 14 November 2014. Available at: https://www.ft.com/content/db052196-6b2a-11e4-be68 -00144feabdc0.

Allee, W.C. & Bowen, E. 1932. Studies in animal aggregations: mass protection against colloidal silver among goldfishes. *Journal of Experimental Zoology*. 61: 185–207.

Anonymous. 1767. *The English Pilot: The Fourth Book. Describing the West Indies Navigation from Hudson's Bay to the River Amazones.* Dublin, Boulter Grierson.

Anonymous. 1955. Royal Society for the Protection of Birds. *Bird Notes* 26: 226.

Anonymous. 1956. Royal Society for the Protection of Birds Sixty-fifth Annual Report (for 1955) 1956, 6-7.

Anonymous. 1968. Mecca for the connoisseur. *Illustrated London News* 252: 22 (8 June).

Anonymous. 2013. Why efforts to bring extinct species back from the dead miss the point. A project to revive long-gone species is a sideshow to the real extinction crisis. *Scientific American*, June 2013.

Armit, I. 2006 Anatomy of an Iron Age roundhouse: the Cnip wheelhouse excavations, Lewis. Society of Antiquaries of Scotland. https://www.socantscot.org/product/anatomy-of-an-iron-age -roundhouse/#:~:text=The%20uniquely%20detailed%20sequence %20at,environment%2C%20and%20with%20their%20neighbours.

Ashley, M. 2016. *The Bird Man's Wife.* Affirm Press.

Audubon, M. & Coues, E. 1897 *Audubon and his Journals* II: 406–411. Scribners, New York.

Bahn, P. G. & Pettitt, E. 2016. *Images of the Ice Age.* Oxford.

Bardason, H. 1986. *Birds of Iceland.* Hjalmar.

Baerends, G. P. 1982. Section V. General Discussion. *Behaviour*, 82: 276–331.

Ben, S., Harvey, M. & Ewing, S. 2023. A history of breeding Slavonian grebes in Britain. *British Birds* 116: 308–318.

Bengtson, S-A. 1984. Breeding ecology and extinction of the Great Auk (*Pinguinus impennis*): anecdotal evidence and conjectures. *Auk* 101: 1–12.

Bent, A. C. 1946. *Life Histories of North American Diving Birds.* Dodd, Mead & Company, New York.

Berland, K. J. 2020. The Passenger Pigeon and the New World myth of plenitude. In Carey, B. *et al. Birds in Eighteenth-Century Literature*, pp. 247–267. Palgrave Macmillan.

Biggar, H. P., 1924. *The Voyages of Jacques Cartier.* Ottawa.

Biggins, J. D., Thompson, J. E. & Birkhead, T. R. 2018. Accurately quantifying the shape of birds' eggs. *Ecology & Evolution* 8: 9728–9738.

Birkhead, T. R. 1977. The effects of habitat and density on breeding success in Common Guillemot (*Uria aalge* Pontopp.). *Journal of Animal Ecology*, 46: 751–764.

Birkhead, T. R. 1978. Behavioural adaptations to high density nesting in the common guillemot *Uria aalge. Animal Behaviour* 26: 321–331.

Birkhead, T. R. 1993. *Great Auk Islands.* T & AD Poyser, London.

Birkhead, T. R. 2016. *The Most Perfect Thing: The Inside (and Outside) of a Birds' Egg.* Bloomsbury, London.

Birkhead, T.R. 2021. The chick-rearing period of the great auk: a mystery solved. *British Birds* 114: 27–37.

Birkhead. T.R. 2022a. *Birds and Us.* Penguin, London.

Birkhead, T. R. 2022b. As rare as hen's teeth: aberrantly coloured eggs of the northern lapwing (*Vanellus vanellus*) and the interface between oology and ornithology. *European Zoological Journal*, 89: 145–159.

Birkhead, T. R. *et al.* 2024. Wooden replicas of great auk eggs created by Stephen James Goodall. *Archives of Natural History.* In press.

Birkhead, T. R., Clugston, D. L. & Fuller, E. 2023. The dispersal of Vivian Vaughan Davies Hewitt's collection of great auk (*Pinguinus impennis*) eggs. *Archives of Natural History* 50: 191–206.

Birkhead, T. R., Fiebig, J., Montgomerie, R. & Schulze-Hagen, K. 2022. The great auk (*Pinguinus impennis*) had two brood patches, not one: confirmation and implications. *Ibis* 164: 494–504.

Birkhead, T. R., Harcourt-Brown, N. & Montgomerie, R. 2023. Great auk gizzard stones. *British Birds* 116: 228-231

Birkhead, T. R. & Nettleship, D. N. 1995. Arctic fox influence on a seabird community in Labrador: a natural experiment. *Wilson Bulletin*, 107: 397–412.

Birkhead, T. R., Russell, D. R. & Thompson, J. E. 2021. Abnormal eggs of the common guillemot *Uria aalge:* the role of stress. *Seabird* 33: 1–17.

Birkhead, T.R., Thompson, J. E. & Biggins, J. D. 2017. Egg shape in common *Uria aalge* and Brünnich's guillemots *U. lomvia:* not a rolling matter. *Journal of Ornithology* 158: 679–685.

Birkhead, T. R., Thompson, J. E. & Montgomerie, R. 2018. The pyriform egg of the Common Murre (*Uria aalge*) is more stable on sloping surfaces. *Auk* 135: 1020–1032.

Birkhead, T. R., Thompson, J. E., Cox, A. R. & Montgomerie, R. 2021. Exceptional variation in the appearance of common murre eggs reveals their potential as identity signals. *Ornithology* 138: 1–13.

Bond, F. 1903. The great auk in art. *Popular Science Monthly.* April 1903.

Bones, M. 1993. The garefowl or great auk. *Hebridean Naturalist* 11: 15–24.

Bourne, W. R. P. 1993. The story of the great auk *Pinguinis impennis.* *Archives of Natural History* 20: 257–278.

Bourne, W. R. P. 1999. History of the great auk: Review of Fuller's *The Great Auk.* *Sea Swallow* 48: 59.

Brodkorb, P. 1960. Great auk and common murre from a Florida midden. *Auk* 77: 342–43.

Buffon, Comte de. 1793. *The Natural History of Birds: from the French of Count de Buffon.* London.

Buffon, J. C. Leclerc de, 1770–1785. *Histoire Naturelle des Oiseaux.* 18 volumes. Paris.

Buxton, J. & Lockley, R. 1950. *Island of Skomer.* Staples, London.

Camphuysen, K. 2022. Mission accomplished: chronic North Sea oil pollution now at acceptable levels, with common guillemots *Uria aalge* as sentinels. *Seabird* 34: 1–32.

Cartwright, G. 1792 (ed. Townsend, C. W. 1911). *Captain Cartwright and his Labrador Journal.* Estes & Co., Boston.

Champley, R. 1864. The Great Auk. *Annals and Magazine of Natural History,* 235-36.

Cheke, A. 2020. The sad exile of the UK's leading ornithological library. *British Birds* 113: 53–54.

Clugston, D., 2013. Obituary: Dr John Alan Gibson (1926–2013). *Scottish Birds* 33: 238.

Clugston, D. 2020. Obituary: David Ronald Wilson (1926–2020). *Scottish Birds* 40: 254–55.

Chilton, G. 2009. *The Curse of the Labrador Duck*. Simon & Schuster, London.

Cole, A. C. 2006. *The Egg Dealers of Great Britain*. Peregrine Books, Leeds.

Cole, A. C. & Trobe, W. M. 2000. *The Egg Collectors of Great Britain and Ireland*. Peregrine Books, Leeds.

Cole, E. 2015. Blown out: the science and enthusiasm of egg collecting in the Oologists' Record 1921–1969. *Journal of Historical Geography* 51: 18–28.

Cook, R. 1993. *The Voyages of Jacques Cartier*. University of Toronto Press, Toronto.

Cosquer, H., Fettu, V. & Bernard, F. 1993. *La Grotte Cosquer. Plongée dans la Préhistoire* [*The Cosquer Cave: Dive into Prehistory*]. Solar, Paris.

Cunningham, E. J. A. *et al.* 2022. The incursion of Highly Pathogenic Avian Influenza (HPAI) into North Atlantic seabird populations: an interim report from the 15th International Seabird Group conference. *Seabird* 34: 67–73.

D'Errico, F. 1994. Birds of the Grotte Cosquer: the great auk and palaeolithic prehistory. *Antiquity* 68: 39–47.

De Lozoya, A. V., Garcia, D. G. & Parish, J. 2016. A great auk for the Sun King. *Archives of Natural History* 43: 41–56.

Eastham, A. & Eastham, M. 1995. Palaeolithic images of the great auk. *Antiquity* 69: 1023–1025.

Egede, H. 1741. *Det Gamle Gronlands Nye Perlustration*. Copenhagen.

Elliot, K. H. *et al.* 2013. High flight costs, but low dive costs, in auks support the biomechanical hypothesis for flightlessness in penguins. *Proceedings of the National Academy of Sciences* 110: 9380–9384.

Fatio, M. V. 1870. *Liste des divers représentants de L'Alca impennis en Europe: oiseaux, squelettes et oeufs. Bullétin de la Societé Ornithologique Suisse* II: 80–85

Fayet, A. L. *et al.* 2021. Local prey shortages drive foraging costs and breeding success in a declining seabird, the Atlantic puffin. *Journal of Animal Ecology* 90: 1152–1164.

Fitzpatrick, A., Bond, J., Büster, L. & Armit, I. 2005. A brief consideration of the later prehistoric appearance and possible significance of the great auk (*Pinguinus impennis*) in the Covesea Caves of north-east Scotland. *The Glasgow Naturalist* 27: 79–82.

Fitzpatrick, J. W. *et al.* 2005. Ivory-billed Woodpecker (*Campephilus principalis*) persists in Continental North America. *Science* 308: 1460–1462.

Flannery, T. 2002. *The Future Eaters*. Grove Press/Atlantic Monthly Press.

Fleming, A. 2024. The last of the great auks: oral history and ritual killings at St Kilda. *Scottish Studies* 40: 28–40.

Fleming, J. 1824. Gleanings of Natural History, during a Voyage along the Coast of Scotland in 1821. *Edinburgh Philosophical Journal* 10: 96–97.

Ford, R., *c.*1968. Romantic histories of some extinct birds and their eggs. Spink & Son Ltd, London.

Frederickson, M. 2010. Action plan for seabirds in Western-Nordic areas. Report from a workshop in Malmö, Sweden, 4–5 May 2010. *TemaNord* 2010: 587, Copenhagen.

Friesen, V. L., Barrett, R. T., Montevecchi, W. A. & Davidson, W. S. 1993. Molecular identification of a backcross between a female common murre, a thick-billed murre hybrid and a male common murre. *Canadian Journal of Zoology* 71: 1474–1477.

Fuller, E. 1999. *The Great Auk*. Southborough.

Fuller, E. 2001. *Extinct Birds*. Cornell University Press.

Gangong, W. F. 1964. *Crucial maps in the Early Cartography and Place-Nomenclature of the Atlantic Coast of Canada*. University of Toronto Press.

Gaskell, J. 2000. *Who killed the great auk?* Oxford University Press, Oxford.

Gaskell, J. 2004. Remarks on the terminology used to describe developmental behaviour among the auks (Alcidae), with particular reference to that of the Great Auk *Pinguinus impennis*. *Ibis* 146: 231–240.

Gaston, A. J. & Jones, I. L. 1998. The Auks, Alcidae. In *Bird Families of the World, Vol. 4*. Oxford University Press, Oxford.

Gibson, J. A. 1984. Scottish Natural History Library. *Society for the History of Natural History Newsletter* 22: 12–13.

Gilbert, W. 2011. Beothuk–European contact in the 16th Century: A re-evaluation of the documentary evidence. *Acadiensis: Journal of the History of the Atlantic Region* 40: 24–44.

Glassford, S. 2006. Seaman, sightseer, story-teller and sage: Aaron Thomas's 1797 *History of Newfoundland. Newfoundland and Labrador Studies*, 21: 1719–1726.

Gould, J. 1832–1837. *The Birds of Europe*. 5 volumes. London.

Greenwood, J. & Greenwood, J. 2018. The origin of the term 'jizz'. *British Birds* 111: 264–274.

Grenfell, W. 1909. *Labrador: The Country and the People*. MacMillan, Norwood.

Grieve, S. 1885. *The Great Auk, or Garefowl (*Alca impennis *Linn.): Its history, archaeology, and remains*. Grange Publishing Works, Edinburgh.

Grigson, G. 1956. A matter of eggs. *Country Life*, 23 August 1956.

Gurney, J. H. 1868. The great auk. *The Zoologist*. Ser 2. v3: 1442–1453.

Haffer, J. 2007. The development of ornithology in central Europe. *Journal of Ornithology* 148 (Supplement 2): S125–S153.

Hammer, J. 2020. *The Falcon Thief: a true tale of adventure, treachery, and the hunt for the perfect bird*. Simon & Schuster.

Harris, M. P. & Birkhead, T. R. 1985. Breeding ecology of the Atlantic Alcidae. In *The Atlantic Alcidae: Evolution, Distribution, and Biology of the Auks Inhabiting the Atlantic Ocean and Adjacent Water Areas* (eds D. N. Nettleship & T. R. Birkhead). Academic Press, London & New York.

Harris, M. P. & Wanless, S. 1990. Moult and autumn colony attendance of auks. *British Birds* 83: 55–56.

Harrison, B. G. 1996. *An Accidental Autobiography*. Houghton Mifflin.

Harvey, M. & Ewing, S. 2023. A history of breeding Slavonian grebes in Britain. *British Birds* 116: 308–318.

Helms, O. 1934. Frederik Faber: an early Danish ornithologist (1796–1828). *Ibis* 76: 723–731.

Hemmings, N., West, M. & Birkhead, T. R. 2012. Causes of hatching failure in endangered birds. *Biology Letters* 8: 964–967.

Hendy, E. W. 1928. *The Lure of Bird Watching*. Jonathan Cape, London.

Hewitson, W. 1831–1838. *British Oology*. Newcastle and London.

Hewitt, V., 1923. An account of the gannets on Grassholm Island. *The Oologists' Record* 3: 70–80.

Hoare, P. 2021. Under the skin of the ocean there's a super-loud discotheque going on. *Guardian*. 9 December 2021.

Hobson, K. A. & Montevecchi, W. A. 1991. Stable isotope determinants of trophic relationships of great auk. *Oecologia* 87: 528–531.

Houlihan, P. F. 1986. *The Birds of Ancient Egypt*. Aris & Phillips.

Houston, A., Wood, J. & Wilkinson, M. 2010. How did the great auk raise its young? *Journal of Evolutionary Biology* 23: 1899–1906.

Hume, J. P. 2006. The history of the Dodo *Raphus cucullatus* and the penguin of Mauritius. *Historical Biology* 18: 69–93.

Hutchinson, R. 2014. *St Kilda: A people's history*. Birlinn, Edinburgh.

Hywel, W. 1973. *Modest Millionaire: Biography of Captain Vivian Hewitt*. Gwasg Gee, Denbigh.

Ingersoll, E. 1923. *Birds in Legend, Fable and Folklore*. Longman, Green & Co., London.

Ingold, P. 1973. *Zur lautlichen Beziehung des Elters zu seinem Küken bei Tordalken (*Alca torda*). Behaviour* 45: 154–190.

Jackson, N. *et al.* 2022. *Lindisfarne: The Holy Island Archaeology Project*. Interim Assessment Report, p. 189. Digventures.

Jardine, W. 1848–1853. *Contributions to Ornithology*. 30–34: 115.

Jenkins, D. & Dunnett, G. 1978. Personalities: Dr W. R. P. Bourne. *British Birds* 20: 123–125.

Jordan, R. H. & Olson, S. L. 1982. First record of the great auk (*Pinguinus impennis*) from Labrador. *Auk* 99: 167–168.

Jourdain, F. C. R. 1934. The skins and eggs of the great auk. *British Birds* 28: 233–234.

Kålund, P. E. K. 1879–1882. *Bidrag til en historisk-topografisk beskrivelse af Island. II. Nord-og Øst-Fjærdingerne*. Gyldendalske Boghandel, Kjøbenhavn.

Kearly, G. 1862. *Links in the Chain*. J. Hogg, London.

Keefe, P. D. 2002. *Empire of Pain: The Secret History of the Sackler Dynasty*. Picador, London.

Keighley, X. *et al.* 2019. Disappearance of Icelandic walruses coincided with Norse settlement. *Molecular Biology and Evolution* 36: 2656–2667.

Langeveld, B. W. 2020. New finds, sites and radiocarbon dates of skeletal remains of the great auk *Pinguinus impennis* from the Netherlands. *Ardea* 108: 5–19.

Lloyd, L. 1854. The great auk still found in Iceland. In *Lloyd's Scandinavian Adventures* II, 495. Bentley, London.

Lockley, R. M. 1957. *Pembrokeshire*. Robert Hale, London.

Lovegrove, R. 1990. *The Kite's Tale; Story of the red kite in Wales*. RSPB, Sandy.

Lucas, F. A. 1890. The expedition to Funk Island, with observations upon the history and anatomy of the Great Auk. *Report of the National Museum 1887–88,* 493–529.

Lyngs, P. 2020. Breeding biology and population dynamics of a colonial seabird: the razorbill. *Dansk Ornitologisk Forenings Tidsskrift* 114: 57–112.

Lysaght, A. M. 1971. *Joseph Banks in Newfoundland and Labrador, 1766: His Diary, Manuscripts and Collections.* Faber, London.

Martin, G. T. 1981. *The Sacred Animal Necropolis at North Saqqara.* Egypt Exploration Society London.

Martin, J. W. R. *et al.* 2000. A new species of large auk from the Pliocene of Belgium. *Oryctos* 3: 53–60.

Martin, M. 1698. *A late voyage to St. Kilda.* London, Gent.

Mayfield, H. F., 1989. In memoriam: Frank Preston. *Auk* 106: 714–717.

McCarthy, M. 2021. *Fergus the Silent.* Privately Published.

McCaughley, D. J. *et al.* 2015. Marine defaunation: animal loss in the global ocean. *Science* 347: 1255641. DOI: 10.1126/science.1255641.

McGowan, B. 2021. Museum egg collections. In Prior, C., *Fragile: Birds, Eggs and Habitats,* pp. 17–19. Merril, London.

Mearns, B. & Mearns, R. 1998. *The Bird Collectors.* Academic Press, London.

Meldgaard, M. 1988. The great auk, *Pinguinus impennis* (L.) in Greenland. *Historical Biology* 1: 145–178.

Montevecchi, W. A. & Tuck, L. M. 1987. *Newfoundland Birds: Exploitation, Study, Conservation.* Nuttall Ornithological Club.

Montgomerie, R. & Birkhead, T. R. (In press.). Estimating the weight of the great auk and its egg. *Ibis.*

Moore, W. 2006. *The Knife Man: Blood, Body-snatching and the Birth of Modern Surgery.* Bantam Press, London.

Morison, S. E. 1971. *The Northern Voyages, A.D. 500–1600.* Oxford University Press, New York.

Morris, P. 2003. *Rowland Ward: Taxidermist to the world.* MPM Publishing, Ascot.

Morris, P. A. 2023. *Taxidermy and the Country House.* MPM Publishing, Ascot.

Moss, S. 2013. *A Bird in the Bush: A Social History of Birdwatching.* Aurum Press.

Murray, J. M. (ed.) 1968. *The Newfoundland Journal of Aaron Thomas 1794.* Longmans, Don Mills, Ontario.

Mynott, J. 2024. *The Story of Nature: A Human History.* Yale University Press, New Haven.

Naumann, J. A. 1795-1817. *Naturgeschichte der Land- und Wasservögel des nördlichen Deutschlands und der angränzenden Länder.*

Naumann, J. F. 1820–1844. *Johann Andreas Naumann's Naturgeschichte der Vögel Deutschlands* ... 12 volumes. Leipzig.

Nettleship, D. N., & Evans, P. G. H. 1985. Distribution and status of the Atlantic Alcidae. In D. N. Nettleship & T. R. Birkhead (eds), *The Atlantic Alcidae: Evolution, Distribution, and Biology of the Auks Inhabiting the Atlantic Ocean and Adjacent Water Areas*, pp. 54–154. Academic Press, London and New York.

Newton, A. 1860. Memoir of the late John Wolley. *Ibis* 2: 172–185.

Newton, A. 1861. Abstract of Mr J. Wolley's researches in Iceland respecting the Gare-fowl or Great Auk (*Alca impennis*, Linn.). *Ibis* 3: 374–399.

Newton, A. 1865. The gare-fowl and its historians. *Natural History Review* 5: 467–488.

Newton, A. 1870. On existing remains of the gare-fowl. *Ibis* 12: 256–261.

Newton, A. 1885. Review of Grieve's *The great auk or garefowl (*Alca impennis, *Linn.), its history, archaeology and remains. Nature* 32: 545–546.

Newton, A. 1896. Great auk. In *A Dictionary of Birds.* A&C Black, London.

Nicholson, M. 1926. *Birds in England.* Chapman & Hall.

Nixon, S. 2022. *Passions for Birds: Science, Sentiment and Sport.* McGill-Queens University, Montreal and Kingston.

Norrgrén, H. 2020. An Alchemist in Greenland: Hans Egede (1686–1758) and Alchemical Practice in the Colony of Hope. *Ambix* 67: 2, 153–173, DOI: 10.1080/00026980.2020.1747305.

Olson, S. 1977. A great auk, *Pinguinis,* from the Pliocene of North Carolina (Aves: Alcidae). *Proceedings of the Biological Society of Washington* 90: 690–697.

Owen, R. 1865. Description of the skeleton of the great auk or garefowl. *Transactions of the Zoological Society of London* 5: 317–335.

Paleczny, M. *et al.* 2015. Population trend of the world's monitored seabirds, 1950–2010 . *PLOS One.* 10(6): e0129342. doi:10.1371/journal.pone.0129342

Pálsson, G. 2024. *The Last of its Kind: The search for the great auk and the discovery of extinction.* Princeton University Press, Princeton.

Parker, G. A. 2021. How soon hath time … A history of two 'seminal'
    publications. *Cells* 10: 287.

Parkin, T. 1911. The great auk: a record of sales of birds and eggs by
    public auction in Great Britain, 1806–1910. *Hastings and East Sussex
    Naturalist* 1: 1–36.

Pastore, R. T. & Story, G. M. 2003. 'Shawnadithit'. In *Dictionary of
    Canadian Biography*, 6. University of Toronto/Université Laval,
    2003–. See http://www.biographi.ca/en/bio/shawnadithit_6E.html.

Pennant, T. 1776. *British Zoology*. London.

Pollard, T. 2021. These spots of excavation tell: using early visitor
    accounts to map the missing graves of Waterloo. *Journal of Conflict
    Archaeology* 16: 75–113.

Pope, P. 2009. Early migratory fishermen and Newfoundland's seabird
    colonies. *Journal of the North Atlantic* 1: 57–74.

Preston, F. W. 1953. The shapes of birds' eggs. *The Auk* 70: 160–182.

Preyer, W. 1862. Ueber *Plautus impennis* Briinn. *Journal für Ornithologie*
    10: 37–356.

Prynne, M. 1963. *Egg-shells: An informal dissertation on birds' eggs in their
    every aspect and also embodying the care and repair of birds' eggs*. Barrie &
    Rockliff, London.

Ray, J. 1678. *The ornithology of Francis Willughby*. Martin, London.

Roberts, A. 1861. Skins and eggs of the great auk. *Zoologist* 1861: 7353.

Rowley, P. 1995. *Chronicles of the Rowleys*. Huntingdon Local History
    Society, Huntingdon.

Salomonsen, F. 1944. The Atlantic Alcidae. *Göteborgs Kungl. Vetenskaps- och
    Vitterhets-Samhälles Handlingar, Sjätte Föliden* Ser. B., Band 3, 5: 1–138.

Schulze-Hagen, K. & Birkhead, T. R. 2023. 'Der fluglose Alk': Johann
    Friedrich Naumann's 1844 account of *Pinguinus impennis* (great auk).
    Archives of Natural History 50: 304–324.

Seebohm, H. 1885. *History of British Birds*. London.

Seebohm, H. 1891. *Coloured Figures of the Eggs of British Birds*. Pawson &
    Brailsford, Sheffield.

Serjeantson, D. 2001. The great auk and the gannet: a prehistoric
    perspective on the extinction of the great auk. *International Journal of
    Osteoarchaeology* 11, 43–55.

Serjeantson, D. 2023. *The Archaeology of Wild Birds in Britain and Ireland*.
    Oxbow, Oxford & Philadelphia.

Sigari, D. *et al.* 2021. Birds and bovids: new parietal engravings at the Romanelli Cave, Apulia. *Antiquity* 95: 1387–1404.

Sigurðsson, G. 1770. Manuscript kept at the Icelandic National Library. Cited in Bardason, 1986.

Smith, E. 2022. *Portable Magic: A history of books and their readers.* Allen Lane, London.

Smith, J. 2006. *Charles Darwin and Victorian Visual Culture.* Cambridge University Press, Cambridge.

Smith, N. A. & Clarke, J. A. 2015. Systematics and evolution of the Pan-Alcidae (Aves, Charadriiformes). *Journal of Avian Biology* 45: 125–140.

Steenstrup, J. J. S. 1885. *Et Bidrag til Geirfuglens,* Alca impennis *Lin., naturalhistorie og saerligt til kundskaben om dens tidligere udbredningskreds. Videnskabelige Meddelelserfraden naturhistoriske Forening i Kjöbenhavn for Aeret 1855,* 3–7: 33–118.

Stresemann, E. & Stresemann, V., 1966. *Die Mauser der Vögel. Journal für Ornithologie* 107 (Sonderheft): 1–448.

Stresemann, E., 1975. *Ornithology: From Aristotle to the Present.* Cambridge, Massachusetts.

Strickland, H. & Melville, A. G. 1848. *The Dodo and its Kindred.* Reeve, Benham & Reeve, London.

Taverner, W. *c.*1718. Captain Taverner's second report relating to Newfoundland. Great Britain, National Archives (PRO), Colonial Office, CO 194/6: 226–241. In Library and Archives Canada, MG 11, Microfilm copy, Reel B-208. Transcription by O. U. Janzen. Available online at http://www2.swgc.mun.ca/nfld history/CO194/TavernorReport2.htm.

Taylor, S. A. *et al.* 2012. Cryptic introgression between murre sister species (*Uria* spp.) in the Pacific low Arctic: frequency, cause, and implications. *Polar Biology* 35: 931–940.

Thomas J. E. *et al.* 2017. An 'aukward' tale: A genetic approach to discover the whereabouts of the last great auks. *Genes* 8: 164. DOI: https://doi.org/10. 3390/genes8060164.

Thomas, J. E. *et al.* 2018. Demographic reconstruction from ancient DNA supports rapid extinction of the great auk. *eLife* 2019, 8:e47509. DOI: https://doi.org/10.7554/eLife.47509.

Thompson, D. W. 1942. *On Growth and Form*. Cambridge University Press, Cambridge.

Thorarensen. J. 1929. *Rauðskinna hin nýrri. (Þjóðsögur, sagnaþættir, þjóðhættir og annálar)*. I. Ísafoldarprentsmiðja Hf, Reykjavík.

Thrush, C. 2016. *Indigenous London: Native Travelers at the Heart of Empire*. Yale University Press, Yale.

Tomkinson, P. M. L. & Tomkinson, J. W. 1966. Eggs of the great auk. *Bulletin of the British Museum (Natural History)* 3: 97–128.

Townsend, C. H. 1930. In memoriam: Frederick Augustus Lucas. *Auk* 47: 147–158.

Tschanz, B. 1990. Adaptations for breeding in Atlantic Alcids. *Netherlands Journal of Zoology* 40: 688–710.

Tuck, J. A. 1976. Ancient people of Port au Choix Newfoundland. *Social and Economic Studies* 17: 262.

Wagner, R. H. 1998. Hidden leks: sexual selection and the clustering of avian territories. *Ornithological Monographs* 49: 123–145.

Walsh, C. T. *et al.* 2001. Social interactions of breeding common murres and a razorbill. *Wilson Bulletin* 113: 449–452.

Walters, J. R. *et al.* 2010. Status of the California condor (*Gymnogyps californianus*) and efforts to achieve its recovery. *Auk* 127: 969–1001.

Whitaker, J. 2021. Recent discovery of erythristic herring gull and great black-backed gull eggs. *The Oologist* 3: 185–208.

Whitbourne, R. 1622. A discourse and discovery of New-Found-Land. In G. T. Cell (ed.), 1981, *Newfoundland Discovered: English Attempts at Colonisation 1610–1630*, pp. 101–206. Hakluyt Society (2nd series), 160. London.

Wilhelm, S. I., Walsh, C. J., Stenhouse, I. J. & Storey, A. E. 2001. A possible common guillemot (*Uria aalge*) x razorbill (*Alca torda*) hybrid. *Atlantic Seabirds* 3: 85–88.

Wilkin, B., Schäfer, R. & Pollard, T. 2023. The real fate of the Waterloo fallen. *Journal of Belgian History* LIII: 8–30.

Wolley, J. & Newton, A. 1864–1907. *Ootheca Wolleyana*. 2 volumes. London.

Wood, J. G. 1862. *The Illustrated Natural History: Birds*. Routledge. London.

Worm, O. 1655. *Museum Wormian*. Elsevier, Leiden.

Index

Page numbers in **bold** indicate figures.

smallpox 159
Smithsonian Institution 124, 125, 165, 185
Snow, David 186
social monogamy 25, 31–32
solitaire, Rodrigues 75
Sotheby's 206, 207
Spink & Son 188, 191, 197–198
Spink, David 188, 191, 200–201, 203, 207
Spink, Dorothy 200
Squire, J. C. 171
St Kilda Club 198
St Kilda, Scotland 33, 37, 53, **62**, 69, 105, 116, 120–121, 140, 198
St Malo Egg 95–96
Steenstrup, Japetus 81–82, 152
Stevens's Auction Rooms, London 92, 151–154, 161–162, 169
Stewart, John 218–219, 220–221
Streymoy, Faroe Islands **62, 227**
Strickland, Hugh 78
Stuart-Baker, Edward 159–160
Stuvitz, Peter 55
swallow, barn 45
Systema Naturae (Linnaeus) 70

Taverner, William 99
Thomas, Aaron 53–54
Thomas, Jessica 219, 220–221
Thompson, D'Arcy Wentworth 100
Thompson, Jamie 101–102
Tinbergen, Niko 30
Tofts Ness, Sanday, Orkney **62, 227**, 228
Tomkinson, John Whitaker 94
Tomkinson, Paule Marie Louise 94
'treading down' of eggs 50, 66, 119–120
Tschanz, Beat 98
Tuglavingaaq 158–159, **158**

Tunnicliffe, Charles 185
turnstone 84
Tvísker, Iceland **62, 227**

Varanger Fjord, Norway 211–212
Vaucher, Alfred 130
Venables, Pat 182, 183–184, 186–187
Versailles, France 68
Victor Emmanuel (ship) 82
Vivian Hewitt's Egg 94, 130, 161, 198, 230, **230**

Wagstaffe, Reg 185
Wallace Hewett's Egg 198, 230, **230**
walrus 61
Ward, Rowland 160, 169, 192
Whatmough, Frank 173, 178, 179
Whitbourne, Richard 51, 52–53, **52**
William Yarrell 97
Williamson, Kenneth 189
Willughby, Francis 68, 70
Wilmot, J. P. 88–89
Wilson, David 183–189, 191–193, 198–199, 200, 202, 204
witches 36, 69–70
Witherby, Harry 168
Wolley, John 77–87, **78**, 88, 212, 223–224
woodpecker, ivory-billed 213, 224
Worm, Ole 68, **69**, 80
wren, Stephens Island 224
Wright brothers 132

Yarrell, William 14
Yarrell's Egg 155, 161, 206, 231, **231**

Zoëga, Geir 82–83, 84
Zoological Society of London 116
Zoologist, The 87–89